Gold in the Americas

◄ **Christopher Columbus's fleet going to conquer the New World**

Nova tipus transacto navigatio novi Orbi by Philoponus Honorius, 1621

Edited by **Hélène Dionne**

GOLD IN THE AMERICAS

Translated by *Joan Irving and Käthe Roth*

SEPTENTRION

MUSÉE DE LA
CIVILISATION
Québec

The exhibition *Gold in the Americas* will be presented in Quebec City from April 30, 2008, to January 11, 2009, and in Paris at the Muséum national d'histoire naturelle from April 4, 2009, to January 19, 2010.

The exhibition *Gold in the Americas* and the present book were produced by the Musée de la civilisation de Québec, under the direction of François Tremblay, Director, Exhibitions department and international affairs, Musée de la civilisation de Québec.

The Musée de la civilisation extends its warmest gratitude to its partners for their commitment, expertise, and financial support in the production of the exhibition *Gold in the Americas*:

The Musée de la civilisation receives funding from the Ministère de la Culture et des Communications du Québec. Les Éditions du Septentrion thanks for their support the Department of Canadian Heritage, the Canada Council for the Arts, and the Société de développement des entreprises culturelles du Québec (SODEC). Les Éditions du Septentrion benefits from the Government of Québec's tax-credit program for book publishing administered by SODEC. The translation was realized with the financial support of the SODEC.

Front cover: Mochica ritual garment in the form of a feline skin, discovered at the Huaca de la Luna site in Peru, Universidad Nacional de Trujillo, Peru, Huaca de la Luna project, Inv. PHLL-56, INC-03, H: 67 cm, W: 31 cm, D: 11 cm. Photograph: Steve Bourget.

Back cover: Spanish gold escudo bearing a cross potent, known as a Jerusalem cross, on its reverse. Late sixteenth century. Photograph: Michael McKeown, The Ulster Museum; Gold Spanish escudo bearing the coat of arms of Philip II. Photograph: Michael McKeown, The Ulster Museum; Portuguese gold coin struck in Lisbon in effigy of St. Vincent (São Vicente), The Ulster Museum, Belfast, Inv. BGR 800. Photograph: Michael McKeown, The Ulster Museum; *Raven Bringing Light to the World*, gold Haida mask by Robert Davidson, diam.: 30 cm. Photograph: © 2007 Royal Canadian Mint – all rights reserved.

Left rabat: Cortés enters Mexico City with great pomp and is warmly greeted by Motecuhzoma. Oil on copper, 1776–1800, Museo de América, Madrid, Inv. 00209. Photograph: Museo de América; "Soleil des Trois-Rivières" monstrance, Claude Boursier, 1663–64, vermeil (silver covered with gold plating), Conseil de la Nation Huronne-Wendat, Wendake, Quebec. Photograph: © Canadian Museum of Civilization, Harry Foster, D2007-12318.

Right rabat: Detail from *Insularum Britannicarum acurata delineatio ex geographicis conatibus Abrahami Ortelii*. In Georg Horn, *Accuratissima orbis delineatio, sive, Geographia vetus, sacra & profana…* Amstelodami: Prostant apud Joannem Janssonium, 1660, Musée de la civilisation, bibliothèque du Séminaire de Québec, Photograph: Idra Labrie, Perspective/Musée de la civilisation.

Publication

Editor: Hélène Dionne
Historical consultant: François Gendron,
   Muséum national d'histoire naturelle de Paris
Bibliographic research: Pierrette Lafond
Iconographic research: Frédérick Bussières
Secretary: Danielle Roy

Exhibition

Director, Historical Exhibitions Department: Pierre Bail
Project manager: Hélène Daneau
Head, Research Committee: José Lopez Arellano
Curators: Annie Beauregard, Nicole Grenier

Publisher

Editorial director: Gilles Herman
Project manager: Sophie Imbeault
Editing: Joan Irving and Käthe Roth
Cover design and layout: Pierre-Louis Cauchon
Index: Roch Côté

© Les éditions du Septentrion
1300, av. Maguire
Sillery (Québec)
G1T 1Z3

Distribution
McGill-Queen's University Press
c/o Georgetown Terminal Warehouses
34 Armstrong Avenue
Georgetown, Ontario
Canada
L7G 4R9

Legal Deposit
Bibliothèque et Archives
nationales du Québec, 2008
ISBN : 978-2-89448-552-1

Contents

Foreword

CLAIRE SIMARD
Executive Director, Musée de la civilisation

Translated by Joan Irving

SYMBOL OF NATIONAL WEALTH, evocation of immense splendours, expression of irresistible power, standard for currencies – gold has exercised its power in every country in the world and influenced the exploration and the colonization of vast regions. Those who have not succumbed to its lure are few and far between. It was the Americas that, from the time of the European conquest to the dawn of the twentieth century, were the world's main suppliers of gold. Even today the production of this precious metal is a challenge that men and women are meeting with skill and ingenuity.

The Musée de la civilisation was compelled to explore this great pan-American adventure in which all of the New World countries have played a part, starting with the pre-Columbian civilizations right up to today's industrial gold. The exhibition *Gold in the Americas* reveals the rich diversity of perceptions and roles of gold, a metal that the earliest inhabitants of the continent called "the sweat of the Sun." Over the centuries, millions of people have dug into the bowels of the Earth, hoping to locate El Dorado and strike it rich. To this day, gold mining represents for the peoples of the Americas a hope that hovers between fortune and misery.

With *Gold in the Americas*, the Musée de la civilisation recounts a truly fascinating story, one that doesn't always look kindly at the protagonists. Aboriginals, conquerors, colonizers, explorers, and entrepreneurs – all had, and still have, a special relationship with this precious metal. At a time when the market value of gold is reaching the stratosphere and the nations of the Americas have started talks on aligning forces, a large exhibition on this theme is right in step with current events.

Gold in the Americas has attracted the interest of numerous collaborators and partners. But above all it has been a very enriching human and professional experience. A project many months in the making, the exhibition was conceived and produced by the Musée de la civilisation, which enjoyed the intellectual support of a team of highly skilled scientists and practitioners evidence of whose work is reflected in the pages of this book. As is its wont, the Musée de la civilisation adopted a global approach for this book, one rooted in the contribution of a multidisciplinary team. All of the authors are respected authorities in their field of expertise, and this book represents a unique opportunity to present readers with their writings on a single theme, *Gold in the Americas*.

It is with pleasure that I acknowledge the remarkable collaboration of the institutions and private collectors who enabled us to put this exhibition together. We received the generous loan of collections from some sixty sources in nine countries: Canada, Colombia, Spain, the United States, France, Ireland, Mexico, Peru, and Portugal. We would also like to acknowledge the important financial and technical contribution of the mining industry as well as the Royal Canadian Mint, who were respectful of the nature of our work every step of the way.

Without this sharing and melding of ideas, know-how, and collections, the exhibition *Gold in the Americas* could never have achieved the scope it has, in its presentation both at the Musée de la civilisation, in Quebec City in 2008, and at the Muséum national d'histoire naturelle, in Paris in 2009.

I hope that for every one of you this voyage to the heart of the Americas is an illuminating and stimulating discovery and that it proves useful in nourishing the important social debates that are now taking place in the Americas and elsewhere.

MUSÉE DE LA CIVILISATION

Québec

BERTRAND-PIERRE GALEY

Executive Director, Muséum national d'histoire naturelle

Translated by Joan Irving

GOLD IN THE AMERICAS, an exhibition conceived and produced by the Musée de la civilisation de Québec, delivers dreams, history, and science all in one place.

Those who conceived the project have put together an all-encompassing exhibition based on the contribution of researchers, curators, and experts from both sides of the Atlantic. The display objects were not assembled simply for the amazement of viewers, but to reveal the great significance of gold in history and in relations among cultures.

The search for this legendary metal unleashed on the New World a wave of men "drunk on a heroic and brutal dream." The often-tragic quest for El Dorado represents the very dark face of an era whose events unfurled in parts of Mexico, Peru, California, Alaska, and also Quebec. But this should not overshadow the ingenuity of the miners and artisans, or the beauty of the work of the goldsmiths who, in accordance with the myths and customs of their cultures, transformed the mineral into precious objects.

Through the exhibition and this publication we also explore the financial role of gold in modern societies, as well as present-day prospecting activities and contemporary geological studies.

In sharing a wondrous and illuminating look at gold – object of nature, object of culture – in sharing an experience through an exhibition, the Muséum and the Musée de la civilisation de Québec are reinforcing a link of cultural cooperation the full meaning of which is achieved by this invitation to the people of France to come and discover the fascinations of *Gold in the Americas*. They will no doubt find it enriching!

Preface
Body of the Americas

Dany Laferrière
Writer

Hélène Dionne does not want us to discuss this very precious subject, gold in the Americas, on the telephone. The gold, she tells me, will soon be arriving in Quebec City. How does one receive this celebrated metal and its multiple metaphors, which contributed so much to the founding of the reality and the imagined space of our continent? Must we dig up that history and its trail of woes? Should we accept that it's just a thing of the past? Who exactly is speaking, here? The heir of the master or of the slave? Because America was born of a crime: the genocide of its indigenous peoples. And one of the reasons for this genocide was the gold. In liquid form, gold ran quicker than heroin in the blood of the conquistadors who, nonetheless, ended up being devoured by anopheles plump with tropical fevers while they lay in their hammocks.

I still bear a few traces of this conquest, and one of them is the siesta. Despite – or because of – the constant buzz of the city, I still take a siesta in the afternoon. After this short telephone conversation, I tuck myself into crisp white sheets, after carefully lowering the blinds. In the soft half-light, which reminds me of the Caribbean, I think about the glinting gold that blinded all of Europe for almost four centuries. According to the Argentine writer Jorge Luis Borges, when you lose your sight, yellow is the last colour to disappear. So I close my eyes and suddenly see all of the gold in the Americas that the Musée de la civilisation has now placed at your feet.

Hélène Dionne is waiting for me, with her smile and her file folders, in a small café on Rue Saint-Denis in Montreal.

– What do you want me to do?

She looks me in the face. Around us, a few customers are quietly reading and drinking their green tea. I always try to find, in the most banal events of daily life, striking signs of the terrors of colonial history. A simple green tea speaks of the aberration known as the West Indies by reminding us of the arrogance of a culture that tried to change the map of the world to avoid acknowledging having sailed the wrong way. India was the other direction, Mister Columbus!

With her sensitive gaze, Hélène Dionne, whose ancestors left Europe for the America that occupied the dreams of Bernardin de Saint-Pierre, the America that was nothing more than a vast work camp scorched by the sun, reaffirms the real meaning of this world, free of both hate and naivety. In the same blink of an eye, I again see my ancestor step out of the hold of a slave ship and tumble, without transition, into the hellhole of Hispaniola. And here we are, face to face in this small Montreal café, the European and the African, two émigrés who will never return to where they started. Have they stayed the same? It's today's America. We should not trust such pure identities in this America, which reels from furious desires and disturbed relationships in these airless nights, embalmed by all the perfumes of its rich flora. Far from Europe, prohibitions fall like flies drunk on heat. And here we are, Hélène Dionne and I, several centuries later, discussing a history that now we share. Near the end of his life, André Breton would sigh, "I'm searching for the gold of times present."

– I'd like you to write the preface for the exhibition catalogue, she says.

And she goes on to explain in detail the project that has had her in a fever for a good while. Gold, it seems, is still ravaging the wild heart of young America. One by one, like nuggets, she presents the authors working on the catalogue. Each has a precise function in the honeycomb where the queen is so carefully collecting her fabulous pollen.

– I find it hard to separate the gold from the body, I say.

◀ *Corps des Amériques*
Photograph: Idra Labrie, Perspective/Musée de la civilisation
Graphic design: Charles St-Gelais/Musée de la civilisation

Carta esatta rappresentante l'Isola di S.Domingo osia Hispaniola, in *Il gazzettiere americano: contenente un distinto ragguaglio di tutte le parti del Nuovo Mondo...* In Livorno: Per Marco Coltellini, 1763.

Musée de la civilisation, bibliothèque du Séminaire de Québec, SQ0025655
Photograph: Idra Labrie, Perspective/Musée de la civilisation

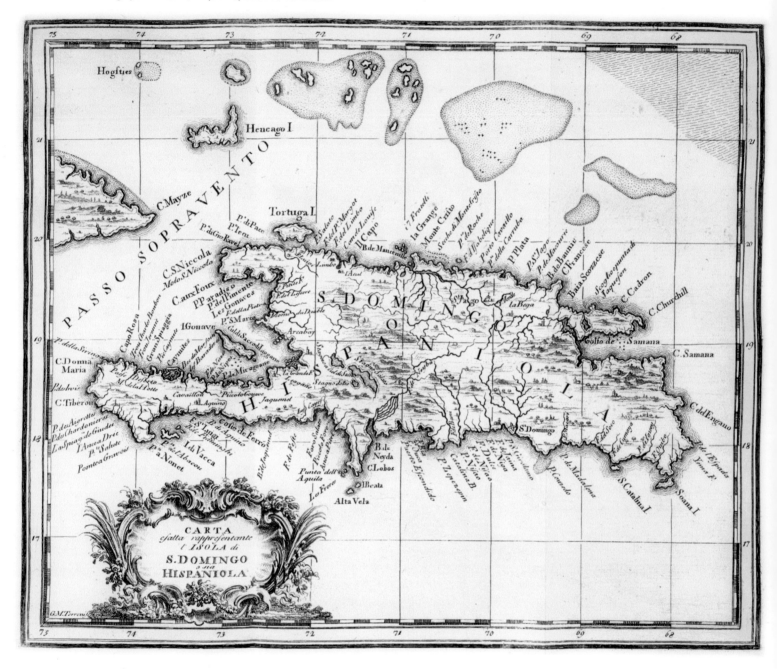

Her eyes begin to shine, but she refrains from commenting, not wanting to influence me. I must not, at any cost, let this already wounded memory become a stolen memory. But that's not what is happening here, so gentle is Hélène Dionne's hand on the still-open wound.

Before going into the museum halls and reading this history of gold, you should know that this is above all a history of men's frantic search for power.

And you should visualize, if possible, the river of blood that flowed because of it. Depending on their place in history, everyone experienced it. For the Indian, it was fast and furious. For the African Negro, it was an endless bloody night. For the European, it was adventure, sickness, and death in a foreign land. Life was different depending on which side of the river of Massacres you were on.

This is how the Haitian poet René Depestre describes the era from the point of view of the Slave:

When the Indian's sweat suddenly ran dry under the Sun,
When the craze for gold drained the last drop of Indian blood in the market
So that not one Indian remained in the vicinity of the gold mines
They turned to the river of muscle in Africa
To supply the next generation of despair
Thus began the rush to the inexhaustible
Treasury of black flesh
Thus began the frenzied scurrying
Toward the shining midday of the black body
And all of Earth rang out with the racket of pickaxes
Amid the thick black ore.[1]

For Depestre, the precious metal was the Negro imported from Africa to take over from the Indian exhausted by sickness and hard labour. The Negro had always viewed this gold as an abomination. And all of America's gold would never weigh as much as the horror of one night in the hold of a slave ship.

So why am I agreeing to introduce this catalogue? Because history is woven from happiness as well as sadness. Because it would be dangerous to start severing it from the things that do not concern us. Because this gold is, after all, here, and it is an incontestable fact. And because it brings with it the energy that sent us hurtling at great speed against the wall of time. It also spawned bloody rituals in which hands drew maps on human skin and traced complex symbols that, over time, were transformed into works of art. Pain that becomes a work of art. We shudder when Depestre evokes "the next generation of despair" – the despair that was passed from the hand of the Indian to that of the Slave, to shine in the night with a sad, yellow glow. We must, of course, spread the net of responsibility to include the Church, which allowed men to be blessed before they were killed.

Seeing the long list of scholars and the diversity of angles from which they have studied this moment in the history of the Americas, I immediately realize that the museum is fully aware of the situation. And that this history is nothing less than the real history of the New World. The saga of gold. Homer noted that the gods endow us with hardships so that we may weave songs from them. These gold objects and the miseries that accompanied them are now in this museum so that men's follies may be seen close up by people who will always seek justice over gold.

For weeks, everywhere I go I carry Hélène Dionne's idea along with me. I do not want the gold to conceal the labour of men and the pain of the women forced to pay for their masters' solitude. And so, to express the coming revolts, I recall the last lines of Depestre's poem:

Let your firedamp ripen
in the secret night of your body
No one will ever again dare
to cast cannons and gold coins
From the black metal of swelling anger.[2]

Detail from *Insularum Britannicarum acurata delineatio ex geographicis conatibus Abrahami Ortelii*, in Georg Horn, *Accuratissima orbis delineatio, sive, Geographia vetus, sacra & profana...* Amstelodami: Prostant apud Joannem Janssonium, 1660.

Musée de la civilisation, bibliothèque du Séminaire de Québec
Photograph: Idra Labrie, Perspective/Musée de la civilisation

---

1. René Depestre, *Minerai noir* (Paris: Présence africaine, 1956).
2. Ibid.

"This land must be desired, discovered, and never deserted."

Letter from Christopher Columbus to King Ferdinand of Aragon, March 4, 1493

Sunrise, September 25, 2005
Photograph: Francis Baillargeon

CHAPTER **1**

THE AMERICAS

The Gold of the Americas

José Lopez Arellano, Ph.D. Antropology, head of the exhibition's Research Committee,

Musée de la civilisation, Quebec City

The Gold of the Americas

JOSÉ LOPEZ ARELLANO

Anthropologist, Head of the Research Committee, Musée de la civilisation, Quebec City

Translated by Käthe Roth

The Americas: Gold and Men

IN THE LATE FIFTEENTH century, the Americas seemed to rise out of the void. The vast stretches of land, the uniqueness of the civilizations, and the variety of resources on the new continents fascinated Europeans and were to radically transform the future of humanity. Gold was the early source of this fascination, and exploitation of gold was the principle that was to fundamentally change the way of life of the indigenous peoples of the Americas and contribute to the development of European colonialism and mercantilism. Dreams of wealth and golden chimeras bewitched the conquistadors, who threw themselves, body and soul, into the first of a long series of gold rushes, unexpected leaps into territorial expansion, through which the New World was reshaped.[1] The endless flow of precious metals toward Europe was to contribute to the construction of a global financial and trade system. If there is a place on Earth whose fate was closely linked to gold, it is the Americas. To bring to light the hidden dimensions of these historic events, we must explore the perceptions, customs, and virtues that the social imagination conferred on gold and the repercussions that human impulses and desires had on the life, culture, and social structure of the peoples of the Americas.

◄ Christopher Columbus lands on Hispaniola (1451–1506).
Engraving from "*De insulis in mare indico super inventis*,"
Basle 1493–94 (Latin translation of Columbus's report to the king of Spain)

Photograph: © Costa/Leemage

A New Continent

In 1491 the Inka ruled the greatest empire on earth.[2] The Europeans who landed on the new continent had no idea of the vastness of the land to be explored or the number of inhabitants, peoples, and cultures that existed there. First contacts with indigenous populations and early explorations were to have a detrimental effect that is still difficult to measure today. The Europeans brought domestic animals with them that carried viruses, including smallpox. "In Europe, the virus was such a constant presence that most adults were immune. . . . Smallpox has an incubation period of about twelve days, during which time sufferers, who may not know they are sick, can infect anyone they meet. . . . Indians had never been exposed to it – they were 'virgin soil,' in epidemiological jargon."[3] The conquistador Cieza de León reported that a great smallpox epidemic had broken out all over the Inca empire, killing more than 200,000 people.

In the sixteenth century, the wealth and power of a nation were measured by its store of precious metals. Each nation therefore wanted to increase its gold reserves at the expense of the others'. At the time, new, emerging economic forces were fighting the influence of the Church, which condemned chrematism, "the art of acquiring wealth," the all-consuming passion that fed the quest for gold (*auris sacra fama*: the damned thirst for gold), and usurious lending. Trade with the East was an unparalleled source of wealth. Europe had long taken advantage of a safe and profitable access route to

---

1. On the consequences of the inflow of American metals on the European economy and on the growth of Spanish mercantilism, see Bernstein, 2000; Chaunu, 1976; and Vilar, 1974. The structuring role of the exploitation of precious metals in the integration of the Americas into the global economy and its consequences for the economy, culture, and development of the former Spanish colonies has been widely documented. Cf. Bakewell, 1989; Brading, 1991, 1977; Galeano, 1981; Serena-Fernandez, 2002; Stein and Stein, 1982.
2. Mann, 2005: 82.
3. Ibid.: 87.

China and India thanks to the hegemony of the Mongolian empire. In 1453, Constantinople fell under the control of the Moslems, who blocked the land route to Asia. To overcome this obstacle, the Portuguese began to explore a route around Africa. It was in this context that Columbus, aware that gold was the greatest prize of all, approached European courts offering his expertise in the search for another route to the East Indies.[4]

Af first, the Spanish Crown rejected the proposal. But the prospects of monopolizing a trade route with the East and reaching the fabulous gold-rich city of Cipango were convincing arguments for a kingdom heavily indebted by the *Reconquista* wars.[5] Columbus made an agreement with the Crown: in exchange for his services and for future riches brought back, he would keep 10 percent of the profits and would be named sole governor and admiral captain of the Indies. In 1492, Columbus set foot on the island of Guanahani, in the Bahamas.[6] Before the stunned eyes of the indigenous peoples, he slashed trees with his sword as a sign of possession. His notary wrote an act of ownership, countersigned by the witnesses present. This first encounter with the "Indians," so called because the Spanish thought that they were in India, marked the beginning of the modern colonial era.

When Columbus returned to Spain, an immense crowd was present for the arrival of the admiral and his fleet. He ostentatiously displayed the heterogeneous group of items that he had brought back from "India": gold, pearls, parrots, and some unhappy, sick Indians, whom he presented to kings as the subjects of the Great Khan. He quickly laid out plans for a much more ambitious expedition involving the foundation of a colony on Hispaniola[7] and exploitation of the gold deposits of the Caribbean islands. During his last voyage, Columbus reached the entrance to the future Panama Canal; he did not find the passage to Asia, but he found gold, a first and fleeting El Dorado. America proved to be extraordinarily rich in precious metals and other treasures.

The Conquest of America

The gold ornaments and objects that the Spanish seized were the product of a great American goldsmithing tradition.[8] In his *Crónica del Peru,*[9] Cieza de León describes the splendours of the Inca empire and the destruction of these treasures by the Spanish. He recounts that the Temple of the Sun at Cuzco had a fountain whose bowl and rim were gold-plated, and that he removed 700 gold plates. The temple garden was adorned with an extraordinary replica of a field in which the ears of corn at the top of golden stems emerged from clods of earth represented by blocks of gold. Twenty golden lamas with their life-size shepherds decorated the wall surrounding the temple.[10] There was much to fuel the conquistadors' greed.

The pre-Hispanic peoples were not uninterested in gold. On the contrary: gold was sought after and valued, but not for its economic function. Pre-Hispanic gold was massively used in the production of personal ornaments and votive objects[11] – emblems of social rank and material supports for a complex religious and shamanic iconography,[12] they accompanied their owner to the tomb. Since gold was sacred, it was found more often than not underground. The sacking of sacred

---

4. According to Columbus's calculations, Earth had a circumference of about 25,500 km. In 1474, Paolo Toscanelli had drawn a map highlighting the proximity between the coast of Portugal and the eastern edge of Asia and calculated a distance of 3,000 nautical miles between the Canary Islands and Cipango (Japan). Columbus shrank this distance to 2,400 nautical miles, arguing that it was possible to reach India in twenty days at sea. The experts of the time had no doubt that Earth was round and that theoretically it was possible to reach China or India by crossing the Atlantic. But the distances calculated by Columbus were, according to the experts, very erroneous. Moreover, no ship of the time could carry all the supplies needed to sail such a distance (Heers, 1991: 121).

5. The *Reconquista* (or Reconquest) was the term for the wars conducted to recapture the kingdoms that had been taken over by the Moslems on the Iberian peninsula. In 1492, Isabella of Castile agreed to Columbus's plan, and he left for the East Indies as the queen's personal agent. The reconquest launched in 718 was completed in 1498. Cf. Stein and Stein, 1982; Thomas, 2003.

6. Historians attribute Columbus's success to the use of new developments in navigation: the compass, the sternpost rudder, and the caravel. Cf. [http://www.cristobal-colon.net].

7. For the jurists of the Spanish court, things were a bit more complicated. Columbus's voyage had transformed the configuration of the world. It was urgent that Spain's taking possession of the new lands be justified legally and that the exclusivity of future discoveries be ensured beyond dispute. Spanish diplomats quickly negotiated with Pope Alexander VI four successive bulls: the Treaty of Tordesillas, an accord that actually divided up the world and signalled the inception of the most powerful colonial empires of the time.

8. Experts agree that pre-Spanish goldsmithery did not have an "economic vocation" but a ritual and emblematic function. Cf. Hosler, 1994; Reichel-Dolmatoff, 1990; Shimada et al., 2000.

9. Cieza de León, 1932: 160 (?).

10. Ibid.

11. Fagan, 1991; Falchetti, 2000; Hosler, 1994.

12. Emmerich, 1965; Reichel-Dolmatoff, 1990; Shimada et al., 2000.

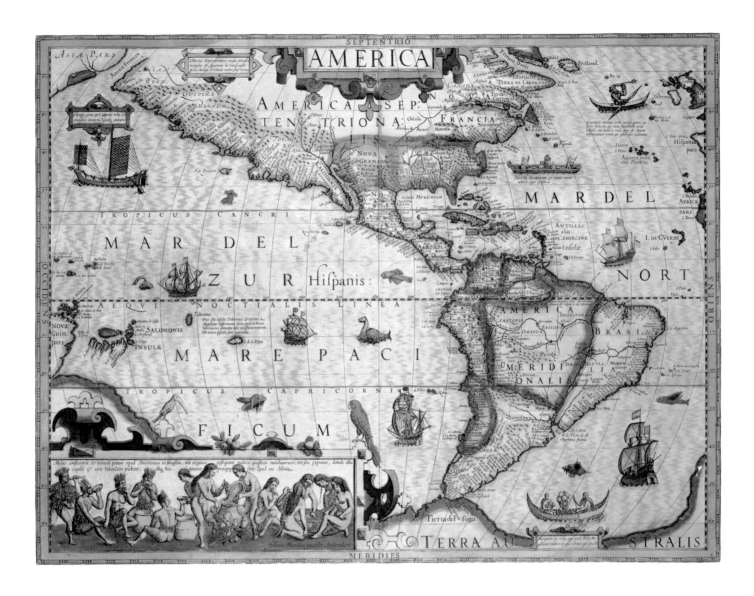

Map of the Americas by Gerardus Mercator, 1595

Photograph: © Stapleton Collection/Corbis

sites and tombs proved to be one of the first forms of appropriation of gold by the Europeans. These magnificent objects were melted into ingots and sent to Seville. Thanks to the consignments from the Americas, the stock of gold in Europe doubled between 1500 and 1650 and tripled between 1650 and 1800.[13]

The Crown's debts were paid in metal, not money. The Spanish Crown decided to accumulate precious metals, and exports of gold were banned – an economic policy known as bullionism.[14] This perception of gold as the materialization of wealth was to have profound repercussions for the economic development of Spain and its colonies.

European Trade Expansion

Given the extent of the New World, the Spanish Crown decided to subdivide the new territories in order to administer them more easily, and it created the viceroyalty of New Spain (1535), followed by the viceroyalties of Peru (1544), New Grenada (1717), and Río de la Plata (1776). The viceroyalty of New Spain alone encompassed today's Mexico, all of Central America as far as Costa Rica, California, Arizona, New Mexico, and Texas.

To make up for the disappearance of the Indians, whose numbers had been decimated by disease and forced labour, the Spanish Crown began to traffic in slaves. Spain did not engage directly in the slave trade, so it required the services of other countries that were involved in this lucrative

---

13. Vilar, 1974: 90–189.

14. Bullionism, developed during the sixteenth century, was one of the first mercantilist theories, and the first mercantilists were called bullionists. These thinkers started with the idea that wealth is based on the abundance of gold and silver, and that these metals must therefore be hoarded. During the sixteenth century, France and England subscribed to bullionism, but with very different results in each case.

market, a practice called *asiento*.[15] Portugal was the first country to gain a monopoly on Spanish *asiento* in America and took charge of capturing and delivering slaves. The English, French, Dutch, Germans, and Danish quickly joined this profitable trade, known as the "triangular trade," one of the cornerstones of European commercial and colonial expansion.[16] The triangular trade consisted of transporting goods – including wheat, wine, and African slaves – from one point to another. The ships left Europe, loaded with trade merchandise, and went to their trading posts established on the coast of Africa, where they traded the merchandise for captives. They then transported the slaves to the American colonies and returned to Europe with holds full of American products. Estimates concerning the number of slaves exported to the Americas vary enormously. It has been calculated that from the early fifteenth century to the early nineteenth century, between 6 million and 50 million people were violently uprooted from African soil to work in the American mines and plantations.[17]

Exploitation of slaves and Indians was the keystone of social, political, and economic life. The abundance of riches "begot festivals and tragedies and made both wine and blood flow."[18] The well-heeled glittered with gold and gems. Religious art and colonial architecture, combined with the radiant glow of gold, were powerful visual and pedagogical tools used to stimulate conversion of indigenous peoples and settlers to the precepts of the Catholic faith. In Europe, the Spanish Catholic empire was buffeted by the rise of Protestantism, the powerful moral renewal that stirred the nascent forces of the modern economy.

The gold ripped from the bowels of American soil did, however, have to make it across the ocean. Spanish convoys loaded with precious metals

Spanish gold ingot produced in Peru between 1600 and 1620

H: 3.8 cm L: 16.5 cm D: 1.6 cm
Weight: 877.7 g
Museo de América, Madrid, Inv.1988/06/14
Photograph: Museo de América

were constantly attacked by pirates.[19] In 1621, the Dutch West India Company obtained a trade monopoly for the Americas; with the creation of the Bank of Amsterdam, the Dutch capital became the international centre for precious metals and home to the world's largest capital market.[20] The essayist Eduardo Galeano wrote, "The Spanish owned the cow but others were drinking its milk,"[21] to describe the twisted paths along which gold transited toward the coffers of the European powers' banks. As gold followed its erratic itinerary, the perception of it began to change: in the eyes of the Dutch bankers, gold was no longer wealth in itself but a vehicle and a source of profits to be treated like any other good. A person who lent gold had the right to a profit, for two reasons: first, because the person ran the risk of losing it; second, because in lending it, the person lost the potential to use it for other transactions – it was thus a loss to be made up for. As circulation expanded, the alchemy of finance was to transform the perception and social use of gold.

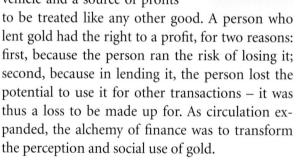

15. *Asiento* was an administrative contract between the Spanish Crown and an individual or company that was engaged in providing a service in exchange for a fee. There were different types of *asiento* for colonial products, but the slave *asiento* was by far the most lucrative. The Spanish *asiento* was tightly linked to operation of the triangular trade. It ended in 1817. Cf. Sanchez Arcilla, 1992.

16. Cf. Thomas, 2006. See also [http://portal.unesco.org/culture/en/ev.php].

17. Empirical data are almost nonexistent and estimates vary from author to author and depend on the sources and the calculation methods. Cf. Thomas, 2006.

18. Galeano, 1981: 35.

19. The United Provinces, at war with Spain since 1566, were determined to challenge the Iberian monopoly over the New World and openly encouraged attacks against Spanish galleons loaded with gold and silver. Dutch pirates were called *vrij buiters* ("free pillagers," which was distorted into *flibustier* in French and *freebooter* in English). This "libertarian" lustre contributed to the trivialization of the crimes committed. With the collapse of Spain and the consolidation of the British empire, the attacks on Spanish galleons became the stuff of fiction filled with heroism and adventure. Cf. Marx, 1992.

20. Most precious metals used by big business came from the Americas, and much of these metals reached Amsterdam. Vilar, 1974: 257.

21. Galeano, 1981: 37.

Dividing Up the World

From the sixteenth to the eighteenth century, the European powers were engaged in a fierce competition to obtain their share of the riches of the Americas. Exploration and colonization were no longer carried out with funds from the royal families' coffers; large trading companies were created to assume these functions. Formed through the issuing of shares, these enterprises aimed ultimately to make their investments profitable. For this to occur, they had to ensure a steady supply of raw materials, guarantee markets for their national industries, control trade routes, and suppress the expansion of competing powers. These companies sought to establish a foothold in the Antilles and secure a place in the triangular trade that was gravitating toward the Caribbean islands and their plantation economy (sugar cane, indigo, and coffee).

In 1607, the London Company created the first English colony, Jamestown, on the James River in what is now Virginia. The company, seeking to make a profit from this undertaking, exerted heavy pressure on the settlers, who went in search of gold and precious stones without the slightest success. To avoid bankruptcy, the company decided to reimburse investors not in cash, but by granting them vast plots of land in the New World.

France put its colony, New France, in the hands of a chartered company, the Company of One Hundred Associates. Champlain and his partners had the mission of discovering a route to China and finding gold. In 1627, the company obtained a monopoly on the fur trade for fifteen years and on all other trade in perpetuity. It agreed to transport 4,000 settlers to New France, assume the administrative costs of the colony, and finance missions to convert the Aboriginals. By entrusting development of the French colony to a private company, the king of France could claim his authority over a vast territory without incurring too high a cost. Chartered companies were to consolidate European mercantile expansion and become the Trojan horse for colonial division of the Americas.[22]

Multiple and complex conflicts of interest among the companies contributed to a long series of intercolonial wars involving the United Provinces, England, and France.[23] In fact, the "reasons of state" for these wars were easily confused with the interests of companies, which did not hesitate to violently quash revolts by settlers or Aboriginals. It is worth noting that it was a trade dispute between New England merchants and the British Crown (the obligation imposed on English settlers to sell their products exclusively in England) that triggered the American War of Independence in 1775.

Gold was a very active, though unseen, factor in these events due to its role in the development of European commodity and securities exchanges.[24] Stock markets enabled charter companies to undertake projects that a single individual could not finance. The securing of American precious metals facilitated the increased control of metal as circulating currency and became a powerful linchpin in the transition toward a modern economy. The English Crown was to follow the example of the Netherlands. In order to increase its control over the flow of money, it put an end to coin striking by hand; the introduction of the first gold guinea, in 1663, marked the advent of mechanized coin striking in England, opening the way to the creation of the Bank of England the following year. The Protestant spirit supported the entrepreneurial culture and the idea that gold brought a "fair" profit. Gold was in the process of transmitting its economic virtues and seductive power to future bank notes, and this was to stimulate the creation of banking and financial institutions.

In America, the rich deposits of ore inspired Spanish and Aboriginal painters, craftsmen, and

---

22. For example, in 1624, the Vereenigde Oostindische, a Dutch company, founded New Amsterdam on the site of today's New York. Sweden, Denmark, Scotland, Russia, and Germany quickly founded their own companies and staked their respective claims to the New World.

23. In North America, the intercolonial conflicts led to a series of profound transformations in the relationships between settlers and Aboriginals. The latent conflict between France and England centred around the fur trade, the fishing zone off Newfoundland, and possession of the Ohio territory, which the Iroquois considered their ancestral land. When the Iroquois came under British protection, England considered itself the legitimate owner of Ohio.

24. Vilar (1974: 249–55) notes that the United Provinces imported gold in ingots, bars, and coins of all sorts, then struck most of it into internationally negotiable coins. In this sense, the Dutch currency became the dollar of its time.

American miners crowded into rail cars,
ready to descend into the mine

Photograph: Courtesy California State Parks, 2008

American Expansion and the Gold Rush

Intercolonial wars totally eroded Spanish imperial authority. England became the strongest global power and grabbed vast territories in North America. The trade dynamism of American merchants, the abundance of natural resources, and exponential population growth facilitated industrialization of the country. The pioneer spirit and optimism of the American people were closely integrated with the ideology of progress, and the purchase or annexation of vast territories around the former colonies (Louisiana, Oregon, Florida, and Texas) legitimized the imminence of imperialistic expansion.

Industrial development provoked intense social crises in the United States and Europe. An economic depression took hold in 1837 and riots in cities became chronic. In 1846, the United States declared war on Mexico; the war ended with American annexation of California, Arizona, New Mexico, and other territories. The victory over Mexico and the announcement of the providential discovery of deposits of gold in California in 1848 confirmed the superiority of the Protestant ethic over Hispanic obscurantism, and thousands of people set out in quest of El Dorado. With new means of travel available, the California gold rush proved to be one of the most powerful mechanisms of American expansion.[26]

The human tide flowed mainly along land routes, following the Oregon Trail, spurred by the motto "Go west, young man, go west!" Poor leaseholders, farm workers, and young adventurers took to the trails in covered wagons, which became the perfect emblem for the conquest of the Far West. Prominent figures, military officers, and wealthy people left New York, Boston, and New Orleans and crossed the Panama isthmus, the

image-makers, who contributed to the creation of colonial Baroque masterpieces. Colourful Hispano-American sacred art acquired brilliance thanks to the application of a layer of gold to paintings and sculptures. The lustre of gilt-covered chapels was a source of jubilation and pride among the faithful. On the other hand, the colonial clergy had to deal with the Americanization of the Miracle: blacks, Indians, and Métis confessed, paid their tithes, and had visions that ended up engendering profoundly syncretic saints and madonnas, such as Our Lady of Guadelupe.[25] Gold never reflected social rank and sacred virtues so brightly as during the Hispano-American colonial era.

---

25. Construction of the Guadelupan myth is another source of debate, but its role in the construction of a national identity is well documented. Cf. Lafaye, 1984.
26. The relationship between the California gold rush and American expansion is not a cause-and-effect one. However, according to Vilar (1974: 424), gold rushes regularly took place in a context of very low prices for all goods and a very high price for gold. In this sense, Columbus's obsessive quest for gold was not coincidental.

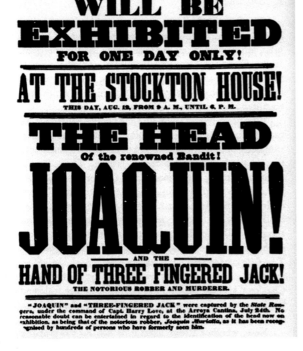

WILL BE EXHIBITED
FOR ONE DAY ONLY!
AT THE STOCKTON HOUSE!
THIS DAY, AUG. 12, FROM 9 A. M., UNTIL 6. P. M.

THE HEAD
Of the renowned Bandit!
JOAQUIN!
— AND THE —
HAND OF THREE FINGERED JACK!
THE NOTORIOUS ROBBER AND MURDERER.

"JOAQUIN" and "THREE-FINGERED JACK" were captured by the State Rangers, under the command of Capt. Harry Love, at the Arroyo Cantina, July 24th. No reasonable doubt can be entertained in regard to the identification of the head now on exhibition, as being that of the notorious robber, Joaquin Murietta, as it has been recognised by hundreds of persons who have formerly seen him.

Poster advertising exhibition of the head
of the renowned bandit Joaquin Murieta, c. 1853
Joaquin Murieta defended the rights of his compatriots,
Mexican, Chilean, and South American miners,
who were threatened by the "gringos," the white
miners, who attacked their claims and their mines.

Photograph: University of Oklahoma Press

shortest sea route to California, though full of danger and exotic adventures. The first voyages were made in clippers, but steamships quickly took over as the means of transport for goods and passengers.

In Europe, it was the *printemps des peuples*. The French revolution of 1848, heavily suppressed, was a dramatic illustration of the social impasses created by the Industrial Revolution. The French government chose to get rid of agitators and unemployed people by holding out the prospect of the riches of Californian gold. A lottery for gold ingots was organized to raise funds and send the outcasts of the revolution to the new El Dorado.[27] The Société des lingots d'or paid to transport more than 3,000 people. They crossed the Atlantic, survived the circumnavigation of Cape Horn, landed in San Francisco, and set out to seek their fortune.[28] In spite of the idyllic images on the advertising posters, the trip to the Californian El Dorado was long, unpredictable, and monotonous. Travellers played cards, got drunk, or wrote down their impressions, giving rise to a veritable industry of accounts of voyages to and adventures in California.[29]

This inflow of newcomers led to extraordinary population growth, mostly men, multi-ethnic, young, and ambitious. In the camps, cupidity, greed, fraud, and racial discrimination were rife. Miners of different nationalities banded together to facilitate prospecting and protect each other.[30] Panning the gold-bearing sand was a thankless and often fruitless job. It was necessary to acquire some expertise.[31] Most prospectors were not out to learn a new trade but to dig up the "big stake" as quickly as possible and then leave.

The veins of the Mother Lode were as rich as had been hoped. In 1848 and 1849, the Forty-Niners[32] extracted 76 tons of gold, and this was only the beginning. Mexican and Chilean miners, heirs to mining traditions with a wealth of expertise, represented a danger to inexperienced American prospectors who were blindly convinced that these riches were theirs by right: conflicts tinged with racism soon exploded. To protect the interests of the new masters of the land, the Californian legislature passed a law in 1850 stipulating that all foreign miners had to purchase a licence costing $20 per month. This was tantamount to a declaration of war.[33] From these conflicts was to emerge one of the most remarkable figures of the gold rush: Joaquin Murrieta, the "Robin Hood of El Dorado,"[34] symbol of the irredentism of the groups violently dispossessed by the forces of progress and the thirst for gold.

27. Cf. Le Bris, 1988; Lemonnier, 1944.

28. The golden chimeras were powerful: no sooner had the ships arrived than they were deserted or simply left to drift in San Francisco Bay. Construction of the dock and expansion of the centre of San Francisco finally covered them up. Their carcasses came back to the light of day during roadwork or construction of new buildings. Cf. Buried Ships: [http://www.sfgenealogy.com/sf/e1852.jpg].

29. Le Bris (1988) reproduces the advertising posters and gives an overview of the most interesting voyage accounts.

30. The Yankees were concentrated mainly at the north end of the Mother Lode. Mexicans, Chileans, Europeans, and Chinese were at the south end. Cf. Le Bris, 1988.

31. Gold in placer deposits was in the form of various sizes of nuggets or crystals rounded by water action; dust, composed of fine particles mixed with sand; and small fragments, needles, or platelets as thick as a nugget. Glittering iron pyrite could easily fool novices. The first instrument used to separate gold from dross was the pan or bowl. It was very difficult to manipulate, and the wash had to be done in a stooped position; stiffness, backaches, and rheumatism were common among gold panners. The pan was quickly replaced by the cradle. Around 1850, the "long tom," a longer, larger version of the cradle, made its appearance. It required teamwork to operate. It was not uncommon to see teams of thirty to fifty miners applied to these tasks.

32. The Forty-Niners was the name given to the first Americans to rush to California. With regard to the volumes of gold put into circulation, Cf. Bernstein, 2000; Caughey, 1975; Le Bris, 1988; Lemonnier, 1944.

33. Mormon Gulch was one of the richest placers in Calaveras Country. It was long exploited by Chileans and Mexicans. After a bloody confrontation, the Chilean and Mexican miners were run off their claims. Cf. Latapie, 2001; Lemonnier, 1944.

34. Cf. Johnson, 2000. The legend of Murrieta is often confused with the legend of El Zorro, a fictional character created in 1919 by Johnston McCulley, whose adventures also took place in the nineteenth century in Hispanic California.

The Gold of the Great Transformation

The Industrial Revolution was the result not of technical inventions, but of social innovations. Indeed, the "great transformation"[36] that created an upheaval in the modern world was made possible by the general acceptance of fictitious goods. It is impossible to understand the contemporary world without virtual money, salaried work, or the growing merchandising of natural elements (earth, water, human intelligence). Thanks to the legitimization of these premises, bank notes were issued in unprecedented numbers. Toward the end of the nineteenth century, North American society was grappling with an over-abundance of dollars, which were constantly losing value. To remedy this situation, a candidate for the presidency of the United States, William McKinley, proposed a reform of the American monetary system in 1897: the abandonment of the bimetallic standard and the adoption of the gold standard as a guarantee of the dollar's value. The struggle between proponents of the bimetallic standard and those of the gold standard was to uncover one of the most insidious aspects of gold in the Western imagination.

In the press, McKinley's supporters declared that there was a natural link between gold, economic stability, prosperity, and progress. For the promoters of the gold standard, gold was the result of the process of natural selection that governs evolution: it was the strongest currency. They declared that gold was naturally reserved for the civilized nations of the white, prosperous race.[37] Silver was closely linked to the characteristics of inferior, backward nations, such as Mexico and China, and, like these nations, silver was essentially corrupt, while gold

The placers were quickly exhausted and so gold-bearing quartz, found buried in mountain slopes, had to be exploited.[35] The former prospectors who strutted about wearing gold chains gave way to miners employed by large companies. Of their desires, dreams, and good or bad luck at El Dorado, there remained only their memories, work tools, and photographs, showing them dazed in the flash of the daguerreotype. The gold that was wrested from the gold-bearing sand, converted into ingots, and purified of human suffering thanks to the foundry owners' stamps and registration numbers, is still sitting in banks' coffers.

The lightning-quick development of the Western metropolises (San Francisco, Sacramento, Los Angeles) and the influence of the American founding myths (the divine mission of the American nation to spread democracy and freedom throughout the world; the conquest of the Far West) confined the ecological upheavals and the very existence of indigenous and Hispanic groups in California to the periphery of social memory. The Mother Lode gold contributed, among other "noble" causes, to funding the American war of secession.

---

35.   The great capital investment facilitated the development of pile hammers, quartz mills, and other infrastructure. The operation of industrialized mines drew many specialized English workers.

36.   *The Great Transformation* is the title of a groundbreaking book by Karl Polanyi, who analyzes the advent of the modern world. He suggests that contemporary society was founded on a utopian postulate: the idea of a market that adapts to itself. This idea arose due to the growth of grand finance, which established an organic link between political organization and the international economic organization. The market acts by reducing everything – human, social, and natural – to its market value. Cf. Polanyi, 1983.

37.   Morse, 2003: 27.

Olympic medals won by
Chantal Petitclerc at the
Athens Paralympic Games, 2004

Photograph: James Duhamel

was naturally pure. The Goldbugs flooded the press with slogans ridiculing the opposition: "Vote for Free Silver and Be Prosperous Like South America." It was a slogan profoundly imbued with racism but very effective.[38] McKinley then orchestrated a major media campaign to launch Americans on the search for a new El Dorado: the Klondike gold rush.[39]

The rush to the Klondike in 1897 was to transform the Canadian West forever. More than 100,000 people went to this particularly inhospitable region to seek their fortune.[40] To reach the deposits, the stampeders had to climb Chilkoot Pass, carrying with them everything they needed to live for a year, about one ton of food and equipment. They then had to travel some 800 kilometres to Dawson City, the true gateway to El Dorado.[41] The future cities of the Pacific coast – Vancouver, Portland, Seattle – were built as the prospectors passed through. McKinley declared war on Spain, seized the Philippines, and, with the strength of the gold from the Yukon and Alaska behind him, established the gold standard as the guarantee of the American dollar in 1900. Colonization of these vast northern territories was carried out to the detriment of the Aboriginals.[42] And, as in California, an unprecedented upheaval of the environment remained as silent testimony of another rush toward El Dorado.[43] The craze ended around 1906, but the United States dollar, spearhead of the new imperial power, ended up dethroning gold as the international currency and Fort Knox became, in the contemporary imagination, the source from which the redoubtable magical virtues of gold would now emanate.[44]

Contemporary Issues

Gold is the emblem of sporting prowess and glory; each country has had a golden century and each athlete who breaks a record earns a gold medal. Each new value is decked out in a "golden" title. Oil is commonly called black gold; rubber was long known as the green gold of South America. Today, drinking water has become not only a planetary issue, but also Canada's blue gold. Gold, a symbol of everything that Western society venerates, is constantly being transformed into goods.

This precious metal is, more than ever, at the core of contemporary social, ethical, and ecological challenges. The organizations that monitor the actions of the large mining companies state that there

38. Ibid. 28–31.
39. In the late nineteenth century, the Hudson's Bay Company essentially ran these territories. In 1880, Joseph Juneau, a French Canadian, left to prospect for the company and found gold at Douglas Island. This discovery led to the foundation of a new city, Juneau, which became the capital of Alaska in 1906. Juneau became the gateway to the Far North.
40. Le Bris, 1988; Morse, 2003.
41. Neufeld and Norris, 1996.
42. Cruikshank, 1991; Morse, 2003: 40; Stone, 1988.
43. Morse, 2003: 101 et seq.
44. The adoption of the gold standard and the replacement of the bimetallic standard in 1870 created the first international monetary system, which rapidly fell into crisis. When the First World War broke out, in 1914, nations began to print more currency without possessing its counterpart in gold. The Bretton Woods Agreements (1946–71) instituted a gold exchange standard system: the U.S. dollar was indexed directly to gold, while other currencies were indexed to the U.S. dollar. These agreements gave rise to the World Bank and the International Monetary Fund. The convertibility of the U.S. dollar into gold ended in 1971 and a system of floating exchanges was instituted in 1973. Cf. Davies, 2002.

is actually more gold in the coffers of banks than in the bowels of Earth.[45] However, the *garimpeiros* of Serra Pelada continue to prospect, Bolivia is experiencing a new gold rush, and exploitation of the gold deposits in the Chilean mountains may cause the Andean glaciers to melt more quickly.

In spite of the legal interdiction against individuals accumulating gold,[46] it is more and more present in daily life due to the new uses to which it is being put in contemporary industry, fashion, and culture. This growing presence is being stimulated by the dematerialization of gold as a currency metal.[47] Moreover, the growing dematerialization of wealth helps to consolidate our feeling of security and our trust in the economic system.[48] Our money is increasingly digital, we can purchase on credit, and we can invest and follow stock-market ups and downs in real time thanks to the Internet. Trust in the system spurs regulation of mining operations: the ecological consequences of metal extraction may be foreseen and neutralized and, even though the health of miners is always at risk, the working conditions are improving.

In countries in the Southern Hemisphere, thousands of miners have no way to earn a living other than by the artisanal exploitation of old mines. In spite of the meagre income that they draw from it, they refuse to abandon it and remain deeply attached to their way of life, stamped with the colonial seal but also with a profound desire for self-determination: "We eat the mines and the mines eat us. For that reason, we have to give rituals to the spirit of the hill so that the hill continue's to reveal the veins of metal to us and so that we can live."[49] The folk miners are not simply vestiges of the past; they are taking part, in their own way, in Western civilization. For them, as for us, gold is the mystery that conveys, through time and space, the tensions and contradictions that run through our relationship with the world and with the Americas.

---

45. Young, 2000: 10–14.
46. The countries that have subscribed since 1933 to the Gold Reserves Act punish the possession of gold. Coin and jewellery collections are exempted. However, this statute has never been strictly enforced. Cf. Davies, 2002.
47. Dematerialization is a recognized legal technique of digital processing of a value and indicates the growing trust that society has in this value.
48. Trust in and a feeling of security about expert systems and other symbolic gauges contribute to the development of delocalization mechanisms (the detachment of social activities from local contexts) and make possible the growth of contemporary modernity. Cf. Giddens, 1994.
49. Nash, 1979: ix.

Bibliography

Bakewell, P. J. 1989. *Minería y sociedad en el México colonial, Zacatecas (1546–1700)*. Mexico City: Fondo de Cultura Económica.

Bernstein, P. L. 2000. *The Power of Gold: The History of an Obsession*. New York: John Wiley.

Brading, D. 1991. *The First America: The Spanish Monarchy, Creoles, Patriots and the Liberal State*. Cambridge: Cambridge University Press.

———. 1977. *Miners and Merchants in Bourbon Mexico, 1763–1810*. Cambridge: Cambridge University Press.

Buried Ships: [http://www.sfgenealogy.com/sf/e1852.jpg].

Caughey, J. W. 1975. *The California Gold Rush*. Berkeley: University of California Press.

Chaunu, P. 1976. *Histoire de l'Amérique latine*. Paris: PUF.

Ciezas de León, P. 1932. *La cronica del Peru*. Madrid: Espasa-Calpe.

Cruikshank, J. 1991. Dän dhá Ts'edenintth'é. *Reading Voices: Oral and Written Interpretations of the Yukon's Past*. Vancouver: Douglas and McIntyre.

Davies, G. 2002. *A History of Money: From Ancient Times to the Present Day*. Cardiff: University of Wales Press.

Emmerich, A. 1965. *Sweat of the Sun and Tears of the Moon: Gold and Silver in Pre-Columbian Art*. New York: Hackert Art Books.

Fagan, B. 1991. *Kingdoms of Gold, Kingdoms of Jade. The Americas Before Columbus*. London: Thames and Hudson.

Falchetti, A. M. 2000. Prehispanic Metallurgy in the Caribbean Lowlands of Colombia. In Collin McEwan, ed., *Precolumbian Gold: Technology, Style and Iconography*. Chicago: Fitzroy, Dearborn.

Galeano, E. 1981. *Les veines ouvertes d'Amérique latine: une contre-histoire*. Paris: Plon.

Giddens, A. 1994. *Les conséquences de la modernité*. Paris: L'Harmattan.

Heers, J. 1991. *1492. La découverte de l'Amérique*. Brussels: Éditons Complexe.

Hosler, D. 1994. *The Sounds and Colors of Power: The Sacred Metallurgical Technology of Ancient West Mexico*. MIT Press.

Johnson, S. L. 2000. *Roaring Camp: The Social World of the California Gold Rush*. New York: W. W. Norton.

Lafaye, J. 1984. *Quetzalcoatl et Guadeloupe: la formation de la conscience nationale au Mexique, 1531–1813*. Paris: Gallimard.

Latapie, D. 2001. *La fabuleuse histoire de la ruée vers l'or. Californie XIXᵉ siècle*. Toulouse: Privat.

Le Bris, M. 1988. *La fièvre de l'or*. Paris: Gallimard.

Lemonnier, L. 1944. *La ruée vers l'or en Californie*. Paris: Gallimard.

Mann, C. C. 2005. *1491: New Revelations of the Americas before Columbus*. New York: Knopf.

Marx, Jennifer. *Pirates and Privateers of the Caribbean*. Malabar, Florida: Krieger, 1992.

Morse, K. 2003. *The Nature of Gold: An Environmental History of the Klondike Gold Rush*. London and Seattle: University of Washington Press.

Thousands of *garimpeiros* (folk miners) climbing the walls of the crater of the gold mine on Serra Pelada, the bald mountain, in Brazil.

Photograph: Stephanie Maze/Corbis

Nash, J. 1979. *We Eat the Mines and the Mines Eat Us.* New York: Columbia University Press.

Neufeld, D., and Norris, F. 1996. *Chilkoot Trail: Heritage Route to the Klondike.* Whitehorse: Yukon Lost Moose Publishing Company and Parks Canada.

Polanyi, K. 1983. *The Great Transformation.* New York: Farrar & Rinehart.

Reichel-Dolmatoff, G. 1990. *Orfebreria y Chamanismo. Un estudio iconografico del Museo del Oro del Banco de la Republica.* Colombia: Ed. Colina.

Sanchez Arcilla, J. 1992. *Las ordenanzas de la Audiencia de Indias (1511–1811).* Madrid: Dykison.

Serena-Fernandez, A. 2002. "Los mecenas de la plata. El respaldo de los Virreyes a la actividad minera colonial." *Revista de Indias,* 60(219): 335–71.

Shimada, I.; Griffin, J. A.; and Gordus, A. 2000. *A Multi-dimensional Analysis of Middle Sicán Objects.* In Colin McEwan, ed., *Precolumbian Gold: Technology, Style and Iconography.* Chicago: Fitzroy, Dearborn Publishers.

Stein, S., and Stein, B. 1982. *La herencia coloniale de América Latina.* Mexico City: Siglo XXI editores.

Stone, T. 1988. *Miner's Justice. Migration, Law and Order in Alaska – Yukon Frontier. 1873–1902.* New York: Peter Lang.

Thomas, H. 2003. *Rivers of Gold: the rise of the Spanish Empire, from Columbus to Magellan.* New York: Random House.

———. 2006. *La traite des Noirs. Histoire du commerce d'esclaves transatlantique. 1440–1870.* Translated by Guillaume Villeneuve. Paris: Robert Laffont.

Vilar, P. 1974. *Or et monnaie dans l'histoire: 1450–1920.* Paris: Flammarion.

Young, J. E. 2000. *Gold: At What Price? The Need for a Public Debate on the Fate of National Gold Reserves.* Mineral Policy Center, Western Organization of Resource Councils.

CHAPTER **2**

THE ROUTES TO GOLD

"Precious but Perilous Things": Gold and Colonialism in Spanish and Portuguese America

DAVIKEN STUDNICKI-GIZBERT, Ph.D. History, McGill University, Montreal

The Myth of El Dorado: Voyagers, Adventurers, and Explorers

CLARA-ISABEL BOTERO, Ph.D. Art History, Director, Museo del Oro, Bogotá, Colombia

Africa and Africans in the Colonial Mining World

DAVIKEN STUDNICKI-GIZBERT, Ph.D. History, McGill University, Montreal

Mukis – Masters of the Andean Mines

DAVIKEN STUDNICKI-GIZBERT, Ph.D. History, McGill University, Montreal

·EST·EFFIGIES·LIGVRIS·MIRANDA· COLVMBI·ANTIPODVM·PR
·QVI·PENETRAVIT·IN· ORBEM·

"Precious but Perilous Things": Gold and Colonialism in Spanish and Portuguese America

DAVIKEN STUDNICKI-GIZBERT

Professor, World History
Department of History, McGill University, Montreal

"Galing gold, wringing rings
precious but perilous things."

Boethius, *The Consolation of Philosophy*

I N ITS MOMENT IT was such a quiet and innocuous scene, indeterminate and open. A ship's boat lowers into the waters, rows through the surf, and makes landfall. On the beach, a small crowd gathers. People from the sea and people from the land join and the exchanges begin: signs and mimes carrying a halting conversation, objects passing hands.

In retrospect, this moment – October 12, 1492, at the beach of Guanahani – has come to be laden with the tremendous weight of history: the history of a continent; the death and birth of entire peoples; the beginnings of modern European colonialism; the creation of the new societies of the Americas.

It is worth noting, since it is more than an incidental detail, that it was gold that came to fix and fatefully direct those first exchanges between the native islanders and Columbus and his crew. During the second day of parleying at Guanahini, Columbus's attentive eye caught a glimpse of what had borne his hopes during the long traverse from Los Palos in Spain. A quick reflection of the

sun's golden light flashed across the beach. There! Among the crowd of islanders, Columbus "saw that some of them wore a little piece of gold hung in a hole that they have in their noses." Having found its focus and lodestone, the conversation suddenly took a more determined turn. What was the source of the gold? "[B]y signs," Columbus would write that evening, "I was able to understand that . . . there was a king who had large vessels of it and had very much gold." Now that gold had come into view, there could be no return. "This land is to be desired, and discovered, and never to be left."[1] Desire, discovery, occupation – the three ascending modes of the colonial process.

Desire

Columbus's unrelenting search for gold forms a constant and central thread through the diaries, logs, and letters that he wrote during those heady years. Native Americans are repeatedly interrogated on the matter and they seem to oblige. Emerging from the strange mix of stilted conversation, hearsay, and fantasy, Columbus hears, or thinks he hears, of islanders heavily laden with thick golden bracelets and ankle bands; an island whose inhabitants harvest the metal at night by the flickering light of their torches; golden kings; and then – as a kind of culminating point in Columbus's fantastic geography of desire – Christendom's most

◄ **Portrait of a man, supposedly Christopher Columbus**

Sebastiano del Pombio, 1519
Metropolitan Museum of Art,
Gift of J. Pierpont Morgan, Inv. 00.18.2
Photograph: © MP/Leemage

---

1. Christopher Columbus, *The Diario of Christopher Columbus's First Voyage to America 1492–1493. Abstracted by Fray Bartolomé de las Casas,* translated by Oliver Dunn and James E. Kelley, Jr. (Norman: University of Oklahoma Press, 1989), p. 71.

fabled land of gold, Ophir, the legendary source of King Solomon's gold, first "discovered" to be in Hispaniola until it slipped away into the great marshlands of the Darien in Panama.[2]

After Columbus came Cortes, Pizarro, Aguirre. After the Spanish came the Portuguese, the English, the French, the Dutch, and others, each questing after the metal, each hitching their projects to the promise offered by the potentially fabulous territories of the New World. Each of these quests recapitulated the successive chapters of conquest, colonization, settlement.

These stories are well enough known. So too, it seems, was the central place that gold occupied within them. Gold was their motivating agent. Gold drove the conquerors who, in their turn, drove the initial cataclysm of colonial expansion. Confronted with the deeds of violence and tribulation, observers at the time – European, African and indigenous – diagnosed an illness, the colonizer's pathological desire for gold. Spaniards, the story ran, suffered from an insatiable appetite that gold alone could quell. Akan merchants on the West African coast felt that this same obsession pushed Dutch and English traders toward the ultimate transgression, a "Metallick idolatry" in which gold was held to be their god.[3] The gold fever never abated, flaring up regularly from the sixteenth century to the present day, driving gold rushes in Brazil, Colombia, Mexico, California, the Yukon, and, today, across the entire breadth of the Americas.

We are usually content to leave it at that, satisfied with a kind of *aura ex machina* in which gold springs upon the scene to set in motion the tragedy of colonial passions and violence. In the early modern period, however, the nature and virtues of gold, as well as the contours and consequences of gold lust, came in for more extensive discussion. European writers knew well that since antiquity, and across the known cultures of humankind, gold stood as the ultimate marker of wealth.[4] Understanding *why* this was so required a closer look at the metal's virtues – that is, its intrinsic qualities – as well as its place within the larger natural and cosmological order. And since virtue could be an active force as much as a quality, it was felt that the peculiar characteristics of gold exerted a particular influence over humans. The virtues of the metal working upon the nature of men generated the passions for gold.

Alonso Barba, the seventeenth-century metallurgist, held gold to be "the most precious of Metals, and the most perfect of all inanimate bodies created by Nature."[5] It was the purest of metals, the most incorruptible, the most resistant to the transformative power of fire. It also had that peculiar quality of ductility: it could be extensively shaped, spread, and flexed without ever breaking the strong bonds that held it together. Finally, gold had curative properties, treating mercury poisoning, dispelling melancholy, consoling the human heart, and even securing perpetual youth.[6] The sixteenth-century natural philosopher Juan de Cardenas explained that these virtues flowed from the privileged relationship that gold held with the Sun. Fertilizing the moist, dark depths of Earth with its rays, the Sun effectively engendered gold, calling it forth from the ground and bequeathing the metal with "its most admirable properties . . . the resplendence, purity and beauty of its rays."[7]

Generative and embryonic theories of the creation of metals held sway

2. Samuel Eliot Morison, *Admiral of the Ocean Sea. A Life of Christopher Columbus.* (Boston: Litttle, Brown and Company, 1942), pp. 35–42.

3. P. de Marees, *Description and Historical Account of the Gold Kingdom of Guinea (1602)* cited in Simon Schaffer, "Golden Means: Assay Instruments and the Geography of Precision in the Guinea Trade," in Marie-Noëlle Bourguet et al. (eds.), *Instruments, Travel and Science: Itineraries of Precision from the Seventeenth to the Twentieth Century* (London and New York: Routledge, 2002), p. 34.

4. José de Acosta, *The Naturall and Morall Historie of the East and West Indies Intreating of the Remarkable Things of Heaven, of the Elements, Mettalls, Plants and Beasts which Are Proper to That Country: Together with the Manners, Ceremonies, Lawes, Governments, and Warres of the Indians,* translated into English by E.G. (1604), pp. 205–06.

5. Alvaro Alonzo Barba, *El Arte de los Metales* [1640], translated by Ross E. Douglass and E. P. Mathewson (New York: John Wiley and Sons, 1923), p. 64.

6. Ibid.; Alfonso X, *Lapidario,* edited by Maria Brey Mariño (Madrid: Editorial Castalia, 1997), p. 64.

7. Juan de Cardenas, *Problemas y Secretos Maravillosos de las Indias* (Mexico, 1591), Coleccion de Incunables Americanos, vol. 9 (Madrid: Ediciones Cultura Hispanica, 1945), pp. 81v, 83v.

The Conquest of Mexico City
Detail of corridor wall along the courtyard of the National Palace of Mexico

Spanish Invaders Attack the Capital with Cannons and Firearms, Diego Rivera, 1929–30
© Banco de Mexico Trust, National Palace, Mexico City DF, Mexico
Photograph: Schalkwijk/Art Resource, New York

across Europe. Interestingly, analogous notions appeared in the Indigenous natural philosophies of the same period. In the Andes Inti Illapa, the divine Sun, also masculine, was the fertilizing agent not only of crops and fields but also of metals that emerged from deep within the Ukhu Pacha, or inner world.[8] The immediate site of conception varied. In certain accounts, venerable frogs laid eggs in the soft folds of the Ukhu Pacha, where they were inseminated by the Sun to create gold. In others, precious metals were generated in a kind of womb-space known as the *mama,* or mother, situated deep within the mountains.[9]

Early modern peoples thus explained the attractions of gold through its powerful virtues and its affiliations with the sun. This power, however, not only fed into the value of the metal, it also generated the perils associated with its extraction and handling. This kind of dualism, the dangerous ambivalence of the qualities of matter, was central to European, indigenous American, and African conceptual systems at the time. If mining was the means through which humans secured the "most precious of all metals," it was also seen as a fundamentally transgressive activity.

The reasons varied. Mining drew humans out of their natural contexts of open land and skies, pulling them deep into the underworld of death and darkness. Mining also artificially accelerated the natural rhythms of metallic creation in which gold was constantly being generated and drawn up to the surface thanks to the attractive agency of the Sun.[10] Since gold was the progeny of the union of Earth and Sun, mining it was a form of theft. Ovid, in a passage often quoted by Renaissance moralists and metallurgists, caught the nature of mining's transgressions:

... greedy mortals, rummaging Nature's store, Digg'd from her entrails first the precious ore; Which next to Hell, the prudent Gods had laid; ... Thus cursed steel, and more accursed gold, Gave mischief birth, and made that mischief bold.[11]

To guard themselves against such mischief, miners from across the Atlantic practised forms of ritual propitiation when they approached the mines. The seventeenth-century Jesuit Anathius Kircher recounted how, in the mercury mines of Hydria, miners would place sets of children's clothing and food on small altars dedicated to the "small demons of the mines."[12] West African miners similarly presented gifts to the spirits of ore beds and observed elaborate taboos. In the Andes, the mine head was ritually soaked in animal blood to assure both the fertility of the veins and the goodwill of the *muki* who guarded them.

But it was once gold was released from the earth that its potential dangers were more broadly realized and a strange kind of dialectic began to operate. The metal of the highest virtues – incorruptibility, nobility, purity – and qualities – the consolation of human hearts – became a catalyst that unshackled humanity's lowest vices. If workers in the mines approached this most powerful of metals warily, taking care to propitiate and appease, others succumbed to

8.   Carmen Salazar-Soler, *Anthropologie des mineurs des Andes* (Paris: L'Harmattan 2002), p. 323.

9.   Jean Berthelot, "L'exploitation des metaux precieux au temps des Incas," *Annales: Economies, Sociétés Civilizations,* Vol. 33, No. 5–6 (Sept.–Dec. 1978), p. 962.

10   Lewis Mumford, *Technics and Civilization* (London: George Routledge & Sons, Ltd., 1934), pp. 69–70; Salazar-Soler, *Anthropologie des Mineurs,* pp. 311–13; Cardenas, *Problemas y Secretos,* pp. 83v–84r. On the idea that the purpose of human *techné* is to perfect nature, see Clarence J. Glacken, *Traces on the Rhodian Shore. Nature and Culture in Western Thought from Ancient Times to the End of the Eighteenth Century* (Berkeley: University of California Press, 1967), pp. 463–67.

11.   Ovid, *Metamorphoses,* Book 1, translated by Sir Samuel Garth, John Dryden, et al.

12.   Salazar-Soler, *Anthropologie des mineurs,* pp. 239–48, 251–52; Marie-Claude Dupré and Bruno Pingon, *Métallurgie et politique en Afrique Central* (Paris: Karthala, 1997), pp. 119–20.

Marine map of the South Atlantic Ocean with parts of South America and Africa.
Pedro Reinel and his son Jorge were well-known Portuguese cartographers who, with the assistance
of Lopo Homen (another cartographer) and António de Holanda (miniaturist) produced
unique and beautiful maps, very accurate for their time.

Pedro Reinel and Lopo Homen, c. 1519. Photograph: British Library/Akg Paris

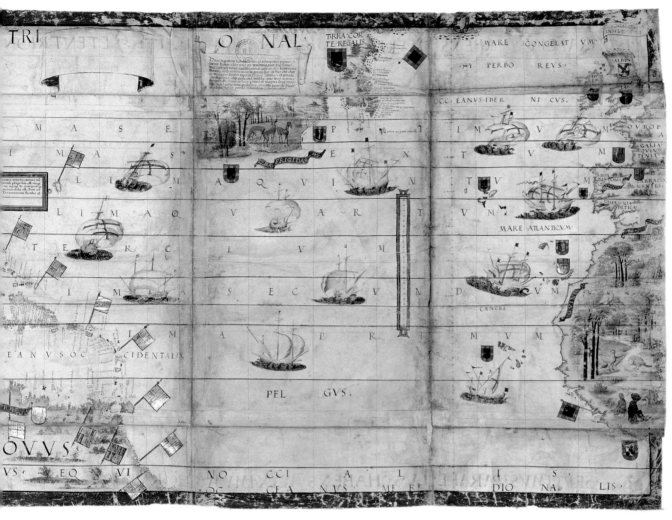

Virgil's famous "*aura sacra fames,*" or sacred hunger for gold, which could make humans lose their reason and corrupt their souls, estrange them from one another, and fracture their communities.

From antiquity to the renaissance, European writers theorized that a susceptibility to desires of various kinds – what Augustine and others described as the passions – was ingrained in the human psyche. Ideally, the higher faculties of reason and intellect held these passions in check. But gold proved to exert exceptional leverage on human desires, perhaps because of its extraordinary virtues, amplifying these desires to the point that the passions ruled over reason and the natural order of things was overturned. This was what Thomas de Mercado had in mind when he wrote of the effects that the advent of gold and other precious metals from the Indies had upon Iberian society. "In times past," he wrote, "men's appetites were moderate but today they are exorbitant and disorderly beyond all measure." The unbounded desire for wealth, which knew "no laws, no measure, no ends," lured people away from their obligations to kin and community.[13]

These theories framed how people in the sixteenth and seventeenth centuries described the motives and actions of the conquerors and colonizers of early America. The Jesuit naturalist José de Acosta – who wrote extensively on the properties and virtues of gold – underscored the corruptive

---

13.   Thomas de Mercado, *Tratos y Contratos de Mercaderes y Tratantes* (Salamanca: Matthias Gast, 1569), pp. 3v, 19v.

Los Mulatos de Esmeraldas

Three Afro-Indian ambassadors from the Republic of Zambo, in the Esmeraldas region of Ecuador, richly adorned with gold, coming to declare their allegiance to the Spanish crown in exchange for their political and social autonomy.

Andrès Sanchez Galque, Indian painter of the Quito painting and sculpture school, 1599

Museo de América, Madrid, Inv. 93-142

Photograph: Museo de América

agency of gold: it blinded the Spaniards, pushing them to unreasoned cruelty and great outrages.[14] For Bartolomé de Las Casas, the metal was the ultimate cause of the "killing and destroying of such an infinite number of [Amerindian] souls."[15] In a famous incident that captured imaginations across Europe, the Jivaro people of the Andean piedmont attempted a symbolic cure of the colonizers' pathological desire for gold. Having seized the Spanish governor of Macas for his depredations in their territories, the Jivaros gathered up the golden loot that he had plundered, melted it down, and then poured it down his throat.[16] As he gasped and writhed, the assembled Jivaros taunted him, "Vomit and spew out the metal which has so unsettled and envenomed both Body and soul, stained and infected your mind's desires . . ."[17]

Discovery

Early modern thinking about gold's virtues and humanity's moral failings made gold-seeking an avenue into a special kind of madness: transgressive, unreasoning, violent. But it was only after 1492 that gold fever took on epidemic proportions as it braided itself into the history of Europe's discovery and occupation of the Americas.

Columbus's search for golden kings and golden lands had met with little success but his reports back to Europe were suffused with an irrepressible conviction that these would indeed be found. His first letter to Ferdinand and Isabel of Spain was immediately set to print, then reprinted nine times in that same year (1493), before going through dozens of re-editions and translations in the following decades.[18] Within a matter of years,

14.   Acosta, *Naturall and Morall Historie*, p. 207.

15.   Bartolomé de Las Casas, *The Devastation of the Indies: A Brief Account*, translated by Herma Briffault (Baltimore: Johns Hopkins University Press, 1992), p. 31.

16.   Michael Harner, *The Jivaro: People of the Sacred Waterfalls* (Berkeley: University of California Press, 1984), p. 21.

17.   Translator's preface to Bartolomé de Las Casas, *A Brief Account of the Destruction of the Indies* (London, 1689). http://www.gutenberg.org/etext/20321.

18.   John H. Elliott, *The Old World and the New, 1492–1650* (New York: Cambridge University Press, 1970), p. 9.

hundreds of colonists made the Atlantic crossing to try their luck at finding fortune in the Indies. They zeroed in on the island of Hispaniola, today Haiti–Santo Domingo, beginning the first gold rush of the Americas.

It began as the outright seizure of the gold objects fashioned by indigenous craftsmen for ornament and ritual. Then followed the placer mining of the gold-bearing alluvial gravels. Indigenous communities were set to forced labour in the rivers and embankments of the island, digging and sifting out the gold with their pans or *bateas*. The Hispaniola gold rush was short-lived – because of the exhaustion of easily exploitable gold deposits and the simultaneous collapse of the local population – but it established a cycle that was repeated in quick succession on the neighbouring islands of Puerto Rico and Cuba.

The Caribbean gold cycle was tremendously important in the early history of the European colonization of the Americas. It formed a prototype, a sequence of phases – exploration, conquest, the coerced mobilization of local peoples into precious metal production, demographic crisis and decline – that was repeated across the mining districts of the continent. It was critical in sharpening European interest in the rich potential of the American territories coming into view. Europe at the time was exceptionally poor in gold reserves and relied on outside sources – especially sub-Saharan West Africa – for what small amounts trickled in. The placers of Hispaniola and the other islands rendered something on the order of 760 kilograms of gold every year from 1502 to 1530, more than double the amount arriving from Africa at the time.[19] The Caribbean experience also worked to confirm late-medieval theories about the accumulation of gold in the equatorial zone. Since the metal was generated by the sun, the theory predicted that gold deposits would be more abundant and numerous around the equator.[20]

The combination of these elements was remarkably powerful: material scarcity, a science that laid out gold across the tropical band, joined with the surge of Caribbean gold in Spain to incite new waves of exploration. Over the course of the sixteenth century, dozens of expeditions were launched toward Mexico, Central America, and South America. Convinced that great gold deposits lay waiting to be uncovered, the leaders of these armed companies relentlessly, violently, interrogated local native Americans about the location of the metal. As they had in the Columbus episode, indigenous informants relayed a mixture of factual and fictitious leads, most often designed to hurry the strangers onward to where they could do no harm. Their stories were then inscribed within a growing body of European writing that construed the new continent as a land of fabulous – indeed, mythical – wealth.[21]

The most famous example of this geographical projection of desire was the myth of El Dorado. The search for the golden kingdom of the lake pulled Spanish conquistadors through the forests of Colombia, Venezuela, Ecuador, the Guayanas, Peru, and Paraguay. With every expedition, the location of El Dorado shifted, slipping across the continent, always eluding them yet also always tantalizingly close at hand – just beyond the next range of mountains or farther up a neighbouring river.[22] As the kingdom of El Dorado moved across the map, it was refracted and transformed into new lands of mythical wealth – the Kingdom of Omaguas, the Land of the Sun, and the Land of Meta – that drew yet other expeditions into the indigenous heartlands of South America.[23] The lure of El Dorado produced a long string of notorious failures – such as Lope de Aguirre and Pedro de Ursua's famously disastrous 1560 expedition from Peru into the Amazon basin – and moments of extraordinary privation. Felipe von Hutten,

19. Earl J. Hamilton, *American Treasure and the Price Revolution in Spain (1501–1650)* (Cambridge: Harvard University Press, 1934), p. 42; Ward Barrett, "World Bullion Flows, 1450–1800," in James Tracy (ed.), *The Rise of Merchant Empires: Long-Distance Trade in the Early Modern World, 1350–1750* (New York: Cambridge University Press, 1990), p. 247.

20 Beatriz Pastor Bodmer, *The Armature of Conquest. Spanish Accounts of the Discovery of America, 1492–1589,* translated by Lydia Longstreth Hunt (Stanford: Stanford University Press, 1992), p. 154.

21. Ibid., pp. 50–51.

22. Juan Gil, *Mitos y utopías del descubrimiento. Tomo III, El Dorado* (Madrid: Allianza Editorial, 1989).

23. Pastor Bodmer, *Armature of Conquest,* p. 161.

Capture of the Inca Atahualpa

Shocked by the surprise attack by Pizarro's men, the Inca king Atahualpa falls from his throne and is captured at Cajamarca in 1531.

In Johann Ludwig Gottfried,
Newe Welt vnd americanische Historien (Frankfurt, 1655)
Musée de la civilisation, bibliothèque du Séminaire de Québec, SQ025651
Photograph: Idra Labrie, Perspective/Musée de la civilisation

for instance, recounted how in 1535 his party was reduced to eating snakes, snails, roots, and, ultimately, the flesh of dogs and humans, which was sold among expedition members against ever-increasing shares of the future treasure of mythical El Dorado.[24]

Not all of the Spanish expeditions searching for the wealth of the Indies ended in failure. Nor were all such fabled lands pure inventions. After their own stint of snake-eating while stranded on the island of Gorgona, Francisco Pizarro and his small party encountered the Inca ruler, Atahualpa, and his army on November 15, 1532, at Cajamarca. Greatly outnumbered, Pizarro chose the gamble of desperation and laid an ambush that, to the surprise of both parties, succeeded in seizing the Inca. With Atahualpa as their hostage, the Spaniards

24. Juan Gustavo Cobo Borda, *Fabulas y leyendas de El Dorado* (Barcelona: Tusquets-Círculo de Lectores, 1987), pp. 30–31.

proceeded to search the Inca army for any treasure. What they found astounded them: some eighty thousand pesos' worth of gold, seven thousand marks of silver, and precious stones. Noting the Spaniards' absorption, Atahualpa detected an opportunity to improve his situation and made Pizarro his famous offer to fill a large room thrice over (once with gold and twice with silver) in exchange for his freedom. The Inca kept his promise but, even as the treasure poured in from the Inca domains, the Spaniards openly discussed his liquidation. Once the ransom was complete, Pizarro proceeded to distribute the treasure – an incredible 1.8 million pesos – among the 168 members of his expedition and the King's royal share, and he then executed Atahualpa.[25]

As in the Caribbean gold cycle, violent seizure was the means through which the first flows of gold moved from the Andes to Europe. Spanish gold-seekers then turned to looting indigenous tombs and shrines for the treasures that they contained. In Colombia, Spaniards and their African slaves moved southward into the Sinú River watershed to excavate the large burial mounds (or *mogotes*) that the Chibchas had built to house their ancestors. At Moche, along the northern coast of Peru, they diverted the local river to hydraulically wash out the gold objects interred in the large Temple of the Sun. A similar attempt was made at Tiwanaku, the

great pre-Inca city on the shores of Lake Titicaca.[26] Like Atahualpa's treasure, these funerary treasures comprised a variety of objects that were systematically melted down, minted, and taxed before being remitted back to Spain. In Quito, the smeltery ledgers of the Royal Treasury reveal that plunder, rather than mining, provided the lion's share of the colony's early gold production.[27]

Occupation

Searching for El Dorados was replaced with prospecting for gold, and it was when the Spanish and Portuguese turned to the organization of mining operations that gold's colonial progression came to its realization. Desire, discovery, and now occupation: the loss of indigenous sovereignty, the installation of new institutions of rule, the subsequent creation of new, colonial, societies that fused elements introduced by the colonizers and the multiple peoples brought together in the crucible

Detail from *Insularum Britannicarum acurata delineatio ex geographicis conatibus Abrahami Ortelii*, in Georg Horn, *Accuratissima orbis delineatio, sive, Geographia vetus, sacra & profana...* Amstelodami: Prostant apud Joannem Janssonium, 1660.

Musée de la civilisation, bibliothèque du Séminaire de Québec
Photograph: Idra Labrie, Perspective/Musée de la civilisation

25. John Hemming, *The Conquest of the Incas* (London: MacMillan, 1970), pp. 48–49 and passim; Silvio Zavala, "Relectura de noticias sobre el botín de los conquistadores del Peru," *Histórica*, Vol. 2, No. 2 (1984), pp. 229–45; James Lockhart, *The Men of Cajamarca: A Social and Biographical Study of the First Conquerors of Peru* (Austin: University of Texas Press, 1972), 13 and passim.
26. Personal communication, Dr. Nicole Couture, Dept. of Anthropology, McGill University.
27. Kris Lane, *Quito 1599. City and Colony in Transition* (Albuquerque: University of New Mexico Press, 2002), pp. 120–22.

of colonial society. Mining is a fundamentally transformative process – one that extracts and refines the raw matter of Earth through processes of motion, heat, and recombination. But so, too, was colonialism – a transformative process of another stripe entirely, bio-political work on the human material of subjected peoples and imported slaves. If the colonial mining districts were one privileged means through which colonial society occupied the territories of the Americas, then their creation, operation, and social life show us the modes and variants taken by this occupation.

As they sought to move past the fables and legends of golden lands, Spanish writers came to appreciate that there was something peculiar about the natural configuration of the American landscape. In Acosta's view, the Americas had been sown with the greatest part of the world's gold and silver. It was, he felt, God's dowry to the continent, a rich endowment guaranteed to attract Christian "suitors" who would carry their civilization and faith to the heathen. Imperial cosmologies aside, Acosta's intuition was correct, for the Americas, it is true, have furnished by far the greatest amount of precious metals to the modern world. In their eighteenth-century heyday, the mines of Spanish and Portuguese America accounted for close to three quarters of the world's gold production.[28]

Since gold deposits were numerous and distributed broadly across the Americas, their discovery and exploitation functioned like so many anchors, fixing colonial society and allowing its early development at multiple points across the territory. After the rivers of Hispaniola, the first mainland zones of gold mining emerged in Veragua, Panama (1507); the Papaloapan and Balsas watersheds in Mexico (1522); the river valleys of the Audiencia de Quito and the Nuevo Reino de Granada (1530s); the rivers and deserts of Chile (1542); and then, somewhat later the northern desert frontier in Mexico (1580s).[29] Then, in the 1690s came the great Brazilian gold rush, as Portuguese slave-raiders and prospectors pushed into the Minas Gerais region west of São Paulo. By the end of the colonial period, gold-mining districts stretched out from Sonora, on the northern Mexican frontier, to southern Chile, from the Pacific coast of Colombia to Brazil.

The extraction of gold was viewed as an artificial acceleration of the normal processes of metallic generation and refinement. The artifice, the application of human labour and technology to the earth, took on two main forms. The most important were the hydraulic techniques used to work gold out of alluvial deposits. Given gold's higher specific density, the application of water and movement allowed the gangue – sands or gravels – to be washed away and the metal to remain. There existed a variety of ways of doing this. The simplest and easiest to organize was for miners to swirl water and the gold-bearing till in large, shallow, funnel-shaped pans (known as *bateas* in Spanish or *bateias* in Portuguese) until the gold dust or nuggets came out. Sluices provided a more elaborate form of the same principle. A long, riffled trough was created – either cut into the ground, or built of wood – into which the placer was deposited. Water was then directed through the sluice, washing away excess gangue and laying bare the gold. In the mines of Carabaya, in the Andean piedmont, indigenous miners built elaborate canalling systems that water painstakingly carved across large granite slabs to the sluices.

There also existed a form of hard-rock gold mining in which miners followed veins of gold underground, burrowing after them in narrow and steeply inclined tunnels. Extensive hard-rock gold mines existed in the Andean towns of Aporoma and Chuquiabo, with multiple shafts and tunnels cutting deep into the mountain.[30] Similar operations were recorded in the Cauca and Magdalena river drainages in Colombia.[31] Miners broke up and dug out the ore-bearing rock with hammers, iron bars, and gads; it was then loaded into large bags weighing some sixty kilograms, which were hauled out by porters. Once out of the shaft, the ore was sent to nearby mills where it was ground up and then the gold extracted through hyrdraulicking or mercury amalgamation. These mines could be quite large. The Espiritú Santo mine on the Darien

28. Barrett, "World Bullion Flows," 225.
29. Robert C. West, *Colonial Placer Mining in Colombia* (Baton Rouge: Louisiana State University Press, 1952), p. 3; Lane, *Quito 1599*, pp. 115–17; Berthelot, "L'exploitation des métaux précieux," p. 948.
30 Berthelot, "L'exploitation des métaux précieux," 955–57.
31. West, *Colonial Placer Mining*, p. 54.

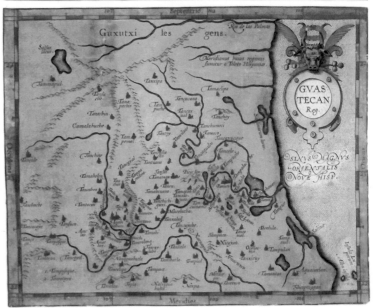

Maps of the gold deposits of Peru,
of the Guastecan region, and of the Florida region.
Map of Peru by Diego Méndez, map of Guastecan by Abraham
Ortello, and map of Florida by Jerónimo Chavez, sixteenth century.

Museo de América, Madrid, Inv. 00410
H: 43.8 cm W: 55.4 cm
Photograph: Museo de América

Peninsula had five levels of galleries, employed roughly two hundred miners, and used mechanical systems to haul out the ore-bearing rock.[32]

Refining gold-bearing ores required extensive manipulation of water and fire, which, in turn, engendered important transformations of the landscapes surrounding the gold districts. The making of canals to direct and control water power at sluices, the creation of large reservoirs to build hydraulic potential even in semi-arid environments, and even the redirection of entire rivers (such as the 1629 diversion of the Rio Nechi in western Colombia) were all ways in which gold-mining operations dramatically reconfigured the hydrology of their locales. At the same time, gold smelting was a highly fuel-hungry process, which, in a pre-electrical and pre-fossil-fuel world,
made important demands on local wood supplies. Observers in Minas Gerais in Brazil noted the disappearance of thick stands of subtropical forests over dozens of leagues surrounding the mines. In northern Mexico, certain operations, such as the Real of Todos Santos, ground to a halt because they had exhausted local supplies of timber.[33] Forests in mining districts tended not to grow back because once they were cleared the land was turned to

32.   Armando Espinosa Baquero, "Datos sobre la explotación y el beneficio de los metales preciosos en Nueva Granada en la época colonial," in *Minería y metalurgia. Intercambio tecnológico y cultural entre América y Europa durante el periodo colonial español* (Seville and Bogotá: Muñoz Moya y Montraveta editores, 1994), p. 493.
33.   Robert C. West, *The Mining Community in Northern New Spain: The Parral Mining District* (Berkeley: University of California Press, 1949), p. 44.

agriculture and pastoralism – food production for the burgeoning populations attracted to the mines. Since many of the mining camps were established in zones previously occupied by nomadic or forest peoples, the spread of fields and pastures in these areas marked a fundamental shift in the landscape triggered by mining.

Colonial gold mining's transformation of the earth was paralleled by processes of social and cultural transformation that took place in the mining camps and towns that sprang up across the gold districts of the Americas. Even at the height of the Brazilian gold rush in the early eighteenth century, the settlements of Minas Gerais were predominantly constellations of small mining camps with only a few towns with populations over five thousand inhabitants.[34] The camps, known as *minas, ranchos*, or *rancherias*, were usually quite small, formed of one to three *cuadrillas* (mining gangs) and their overseers – from fifteen to twenty men and women. Large camps were composed of up to 150 people.[35] What the camps and towns lacked in

34. Charles Boxer, *The Golden Age of Brazil, 1695–1750: Growing Pains of a Colonial Society* (Berkeley: University of California Press, 1964), pp. 51–53; Kathleen J. Higgins, *"Licentious Liberty" in a Brazilian Gold-Mining Region: Slavery, Gender, and Social Control in Eighteenth-Century Sabará, Minas* Gerais (University Park, PA: Pennsylvania State University Press, 1999), p. 19.

35. There were larger mining camps, but even these were never larger than 150 individuals. West, *Colonial Placer Mining*, pp. 102–03; Higgins, *"Licentious Liberty,"* pp. 47–50; 60–61.

size, they made up for in numbers: hundreds certainly, and perhaps thousands, of them came into being over the course of the colonial period.

During the bonanza years that followed the first successful strikes, these communities worked like vortices, sucking in people at a tremendous rate. Within weeks and months, dozens of camps were set up around the deposits. At Ciengeguilla, in the desolate Altar desert of Sonora, 1,500 miners appeared within four months of first strike; eight months later, they numbered 5,000.[36] The gold came out in great quantities, giving the camps a febrile and unruly cast. A lifetime's fortune was made in a single panning; gambling, drinking, theft, and violence set the camps off as places on the extreme edge of colonial society. The indigenous and African workers pressed into working the deposits felt this edge most keenly of all. In bonanza times, the greatest determinant of getting the gold out of the ground was labour. This led to an almost desperate scramble by miners to conscript workers. Africans, already commodified by the workings of the Atlantic slave system, could be purchased and set to work. Indigenous peoples, in theory, were protected from enslavement by the Spanish and Portuguese crowns, but along the gold frontier the rush for labour left no room for legal niceties. Raiding for "rebellious" Indians became standard practice in the Mexican north. Enslavement, impressment, the use of violence, and prison-like forms of control to discipline workers were forms of social transformation that turned certain humans into the tools of others.

Yet transformation could be creative, too. In this respect, the camps and towns of the gold districts were highly cosmopolitan and culturally complex communities. They drew in people from all over the Atlantic world. The districts of Colombia and Brazil were dominated by the different nations of Africans brought over as slaves. In Sabará, Minas Gerais, dozens of languages, including Portuguese and various Creoles, could be heard in the riverbeds and camps. In Mexico, mining camps had a more indigenous cast, but the same general pattern of cultural multiplicity held here too. The workforce at the gold mines of San Pedro, near San Luis Potosí, Mexico, was composed of no fewer than six distinct indigenous groups, to which were added African and Afro-Mexican miners and overseers, and Europeans, who occupied the full range of positions from mine worker to priest to proprietor. The composite nature of the mining settlements gave rise to patterns of cultural fusion and recombination characteristic of the larger phenomenon of colonial *mestizaje*.

The cultural dynamics of *mestizaje* contoured and cut against the colonial program of cultural acculturation that sought to turn indigenous and African peoples into Christian subjects. There was much about the gold districts that escaped the full control of colonial authorities. With the exception of certain small towns that were formed as the administrative centres for a given district, gold-mining camps were ephemeral, moving on to new claims once the gold deposits began to be exhausted. The overall image, and one that characterized gold-mining districts across the continent, was of a constantly shifting and dispersed population of miners. In the late eighteenth century, the newly appointed governor of Antioquia in Nueva Granada was incapable of assessing how many active mining operations existed in his jurisdiction. Posing the question to local officials, he found that, although they had lived and worked there their entire lives, they were incapable of rendering an accurate count. The movement of the miners and the extent and difficulties of the terrain resisted all attempts at state observation.[37]

Much the same can be said for the authorities' efforts at controlling the circulation of gold in their territories. Gold's informal economy came into being in the very streambeds, sluice boxes, or mineshafts where it was first extracted. Nuggets and dust vanished onto the bodies of workers, who would then make the gold reappear in small smelting ovens operated by their fellows. In Colombia and Brazil, these hidden streams of gold were used to secure manumission for thousands of African slaves.[38] Just as African and indigenous workers slipped gold past the eyes of the mine owners, so, too, did miners and merchants slip the metal past the gaze of the colonial state, anxious to appropriate

36. Robert C. West, *Sonora: Its Geographical Personality* (Austin: University of Texas Press, 1993), p. 85.

37. Ann Twinam, *Miners, Merchants, and Farmers in Colonial Colombia* (Austin: University of Texas Press, 1982), pp. 18–19.

38. West, *Colonial Placer Mining*, p. 88; Higgins, "*Licentious Liberty*," pp. 39–40, 80–81, 151–159.

their own stream of wealth coursing from the districts of the Americas. Gold, like precious stones and pearls, was a great favourite of contrabandists. Huge amounts of unregistered gold moved from Minas Gerais to Brazil's Atlantic ports. The dense forests and the complex and shifting networks of backwoods trails made smuggling child's play for the veteran woodsmen who dominated this traffic.[39] Nor was the unregistered circulation of gold purely the work of professionals. Gold was slipped out of Minas Gerais and into the Atlantic economy by people of all stations by any means possible: in barrels and sugar chests and even inside hollowed timbers of the ships making the Atlantic run.[40] Priests took advantage of their clerical immunity to pass by state officials, often concealing gold dust in hollowed-out wooden images of saints.[41]

Connections

Once free, gold flowed into Europe and across the world economy. Historians estimate that between 2,500 and 2,700 tonnes of the metal was produced by the mines of the Americas, of which between 50 and 60 percent was drawn out overseas. The impacts of American gold on the European economy in the early modern period were notable. The rising inflows of the metal had the effect of helping monetize the European economy. This process had begun in the fifteenth century, with the arrival of West African gold through the Portuguese trade, but it accelerated over the centuries as mining operations in the Americas were established. By the end of the eighteenth century, the net effect of this influx was to make Europe one of the regions with the highest per capita levels of gold stock in the world.[42] More important, perhaps, than the total amount of gold being pooled in Europe was the manner in which the increasing amounts of

bullion served to accelerate the velocity of monetary circulation. Gold, and silver, provided the raw capital that credit instruments leveraged and set into motion through commercial exchanges. This, in turn, helped accelerate the cadence with which such transactions could take place and thus served to increase the total amount of capital effectively in circulation in the European economy.[43]

This essay began with Columbus and it ends with Bartolomé de Las Casas, who also wrote reams about the place of gold in the first chapter of European – that is Luso-Iberian – colonization in the Americas. In *De Thesauris* (Of Treasure), a text written immediately before his death at the age of eighty-two, the Dominican elaborated a kind of codicil on the theme of gold and colonialism. He returned to the conqueror's practice of raiding tombs for treasure. Not only was plundering the tombs a matter of theft, it also violated deeper natural and moral laws that protected burial grounds and their contents from violation.[44] Las Casas then amplified the legal and moral injunction against grave-robbing to cover all forms of wealth obtained from Indigenous lands. Like tomb raiding, gold mining was ultimately a form of theft, since neither the Spanish monarchs or their subjects in the Americas had a legitimate claim on indigenous peoples and their territories.

Las Casas's argument led him to a broad and distributive theory of colonial implication. Colonialism's theft, in his view, was organized by a system that implicated not only the conquistadors but also the merchants who outfitted them, the ship's captains who transported them, and even the tailors who furnished their clothes.[45] The value represented by gold had coursed through these lines of relationship, but so too, then, did the responsibilities for the mischief it had wrought.

---

39. Boxer, *The Golden Age of Brazil*, pp. 43–44.

40   Ibid., 201.

41. Ibid., 54; A. J. R. Russell-Wood, "Clerical Participation in the Flow of Bullion from Brazil to Portugal during the Reign of Dom João IV (1706–1750)," *Hispanic American Historical Review*, Vol. 80, No. 4 (2000), pp. 815–37.

42. Barrett, "World Bullion Flows," 253.

43. Harry A. Miskimin, *The Economy of Later Renaissance Europe* (New York: Cambridge University Press, 1977), pp. 36–37.

44. Fray Bartolomé de Las Casas, *Obras completas. Tomo 11.1 De Thesauris*, translated and edited by Angel Losada (Madrid: Allianza Editorial, 1992), pp. 47, 49.

45. Las Casas, *De Thesauris*, pp. 507, 509, 511; Henry R. Wagner and Helen R. Parish, *The Life and Writings of Bartolomé de Las Casas* (Albuquerque: University of New Mexico Press, 1967), pp. 234–35; Marcel Bataillon, "Les 'douze questions' péruviennes résolues par Las Casas," in Marcel Bataillon (ed.), *Études sur Bartolomé de Las Casas* (Paris: Centre de Recherches de l'institut d'Études Hispaniques, 1965), p. 260.

The Myth of El Dorado: Voyagers, Adventurers, and Explorers

Clara-Isabel Botero

Director, Museo del Oro (Bogotá, Colombia)

Translated by Käthe Roth, from the French translation by Louis Jolicœur

Engraving portraying the Legend of the Golden Indian
Qvomodo Imperator Regni Gvianae, Nobiles Svds or. In *Americae pars VIII: continens primo, descriptionem trium itinerum nobilissimi et portissimi equitis Francisci Draken.* Francofurti: Impressae per Mattheaum Becker, 1599, pl. XV

Musée de la civilisation, bibliothèque du Séminaire de Québec, SQ004263
Photograph: Idra Labrie, Perspective/Musée de la civilisation

THE MYTH OF El Dorado, the source of a myriad of legends and stories in Latin America, originated in the accounts of conquistadors and European chroniclers of the fifteenth, sixteenth, and seventeenth centuries. The idea of the "Golden Indian" appeared very early in Gonzalo Fernández de Oviedo's *Historia General y Natural de las Indias,* in which he referred to the existence of "a man or a prince who applies a very aromatic gum or liquor

Aerial view of Lake Guatavita
Lake Guatavita is where the ritual of the Golden Indian took place. According to the legend, the Indian, his body covered with gold powder, jumped into the water with a number of gold objects as offerings to the sun god. Knowing what lay at the bottom of the lake, a number of adventurers and, later, companies, tried, over the centuries, to drain the lake or lower the water level in order to recover these fabulous objects.

Photograph: Hans-Jürgen Burkhard/Bilderberg Archiv fotografen

consecration of the future cacique, as related by the chronicler Juan Rodríguez Freile in 1636:

In this lagoon, a large raft made of rushes, very ostentatiously ornamented, was made. . . . The heir was undressed, coated with a sticky substance, and covered with gold powder so that he was completely covered with the metal. The golden Indian made an offering of all the gold that he was carrying, by throwing himself into the middle of the lagoon . . . and when the raft returned to land, the festival started with much dancing and celebration. The ceremony concluded with the recognition of the Indian as the new prince. It was from this ceremony that came the much-vaunted name El Dorado, which was to cost so many lives and fortunes.[4]

to cover himself with gold powder from head to foot, his entire body as resplendent as a piece of gold handworked by a master craftsman."[1] This myth, built up by European chroniclers, inspired expeditions conducted by voyagers, adventurers, and explorers from the sixteenth to the twentieth century. In search of El Dorado, hidden in secret lagoons where offerings ceremonies took place, fortune-hunters scoured the eastern cordillera of Colombia.

The legend was recounted in great detail in Fray Pedro Simón's *Noticias Historiales de las Conquistas*, written in 1621 and 1623. Simón reprised Juan de Castellanos's account of a Quito Indian and his chief who apparently mentioned the custom of offering sacrifices in the lagoons.[2] Simón names the Guatavita lagoon, which he describes as "the most famous and most often visited temple."[3] This legend was converted into an account of the

Many Europeans suddenly became interested in the gold in the Guatavita lagoon. According to the account of the first explorer, Lázaro Fonte, captain for Gonzalo Jiménez de Quesada, the latter did not want to leave with empty hands. Hernán Pérez de Quesada was said to have lowered the water level by three metres in order to recover between 3,000 and 4,000 gold *castellanos* (between 13.8 and 18.4 kilograms).[5] The first attempt to recover precious objects from the Guatavita lagoon, authorized by the Crown, had been made by Antonio de Sepúlveda in 1562. Endowed with a royal order stipulating that he should make these explorations at his own expense and risk, he got to work.[6]

A veil of silence then covered the pre-Hispanic vestiges until the nineteenth century. In 1820, the newly independent Republic of Colombia allowed many foreign visitors and scientists to visit the Guatavita lagoon. Visitors floated on a raft while an expert related mythic and poetic accounts drawn from texts of chroniclers describing the consecration ceremony for the Muisca cacique.[7] Scientists

1. Gonzalo Fernández de Oviedo, *Historia General y Natural de las Indias*, vol. 4, Cap II (Madrid, 1855), p. 383.
2. Pedro Simón, *Noticias Historiales de las Conquistas de Tierra Firme en las Indias Occidentales*, vol. 3 (Bogotá: Tercera Noticia Historial, 1981), pp. 320–21.
3. Ibid., p. 323.
4. Juan Rodríguez Freile, *El Carnero* (1636) (Bogotá, 1994), pp. 80–81.
5. Roberto Lleras, "Las Ofrendas Muiscas en la Laguna de Guatavita," in *El Mar, Eterno Retorno* (Bogotá, 1998), p. 23.
6. "Capitulación que se tomó con Antonio de Sepúlveda sobre la Laguna de Guatavita y del montecillo della Año de 1562," in Antonio Basilio Cuervo, *Colección de Documentos Inéditos sobre la Geografía y la Historia de Colombia*, vol. 4 (Bogotá, 1894), pp. 135–36.
7. Charles Stuart Cochrane, *Journal of a Residence and Travels in Colombia during the years 1823 and 1824* (London, 1825), pp. 199, 205; John Potter Hamilton, *Travels by the Interior Provinces of Colombia*, p. 128.

were not the only ones interested in the lagoon; the colonial dream of treasures hidden in burial and offerings sites of the pre-Columbian peoples was still alive. An English captain and the vice-president of the Republic, partners in a trading company, made another, unsuccessful attempt to empty the Guatavita lagoon once again in 1823–24.

Around 1850, interest moved to another sacred location, the Siecha lagoon, and in 1856, a private Colombian company succeeded in emptying it. Uncovered were some emeralds and gold objects,[8] among which was one absolutely extraordinary piece: a gold raft representing the religious ceremony of consecration of a Muisca cacique surrounded by his entourage.[9] The raft was initially acquired by a private collector in Bogotá, Gonzalo Ramos Ruiz. The Berlin Museum of Ethnography, interested in the unusual object, asked the German ambassador in Bogotá to purchase it.[10] However, Ramos refused to sell it to the museum and offered it instead to the wife of Salomón Koppel, a German banker living in Bogotá. The director of the museum in Berlin, the ethnologist Adolf Bastian, wrote to Koppel to advise him that this "fabulous object" would be a highly appreciated gift for the emperor, and that it would be much better put to use in the service of science in the Berlin public royal collection than in a private house in Bogotá.

In 1883, the Siecha raft, "one of the most important and interesting gold antiquities of the Chibchas" according to Edouard Seler, curator of American collections at the Berlin museum, finally arrived in Germany. There was a long exchange of letters between Bastian and Koppel before the latter agreed that the Chibcha raft should be shipped to Berlin. Soon after it arrived in Bremen, however, a huge fire laid waste to the port's warehouses, and it was burned.[11]

A lithograph of the raft was published in *Kultur und Industrie Südamerikanischer Völker* in 1889. According to the authors, the lithograph portrays a reproduction in gold-plated silver, made from "the gold original from the Siecha lagoon, which weighed 162 grams and belonged to Koppel, and which was lost in transit to the Royal Museum of Berlin."[12]

In 1897, the idea of emptying the Guatavita lagoon re-emerged. A small company in Bogotá made the attempt, but failed. Contractors Limited, an English company founded in 1900, obtained the rights from the Colombian company. In 1904, the lagoon was finally emptied.[13] The company described what it found:

Laguna de Siecha, Cundinamarca.

Lithograph of the Siecha raft
This magnificent object was destroyed in a fire when it arrived at the port of Bremen in 1889.

Image published in Max Uhle, Alphons Stübel, Wilhem Reiss, and Bendix Koppel, *Kultur und Industrie Südamerikanischer Völker*, 2 Vols, Vol. 1, Ilustration 21. Leipzig, 1889.

Photograph: Biblioteca Luis Angel Arango, Bogotá, Colombia

8. Joaquin y Bernardino de Tovar, "Reflexiones sobre el Dorado y su Descubrimiento," *La América*, No. 76 [Bogotá,] (9 April 1873), p. 303.

9. Liborio Zerda, "Antiguedades Indígenas," *Anales de la Universidad*, No. 61 [Bogotá] (1874), p. 186.

10. Harassowitz, secretary at the German embassy in Bogotá, to Adolf Bastian, Bogotá, 7 Nov. 1876, in *Acta de los Viajes del Director Profesor Dr. Bastian, 1875–1880, Pars IB 10*, Archives of the Berlin Ethnographic Musum.

11. Eduard Seler to Carl Fluhr, Berlin, January 3, 1899, *Acta America*, vol. 19 (May 1898–July 1899), 1369/98. AMVB and Eduard Seler, "Ueber Goldfunde aus Kolumbien," in *Congrès international des Américanistes, compte rendu de la Dixième Session* (Stockholm, 1894; Stockholm, 1897), p. 65.

12. Max Uhle, Alphons Stübel, Wilhem Reiss, and Bendix Koppel, *Kultur und Industrie Südamerikanischer Völker*, 2 vols. (Leipzig 1889), vol. 1, il. 21.

13. *Description and details of articles recovered from the Sacred Lake of Guatavita, Republic of Colombia, South America, through the operations of Contractors, Ltd.*, London, 1912, p. 5.

Gold Muisca boat

Museo del Oro, Inv. M.O.11.373

Photograph: Juan Mayr/Museo del Oro

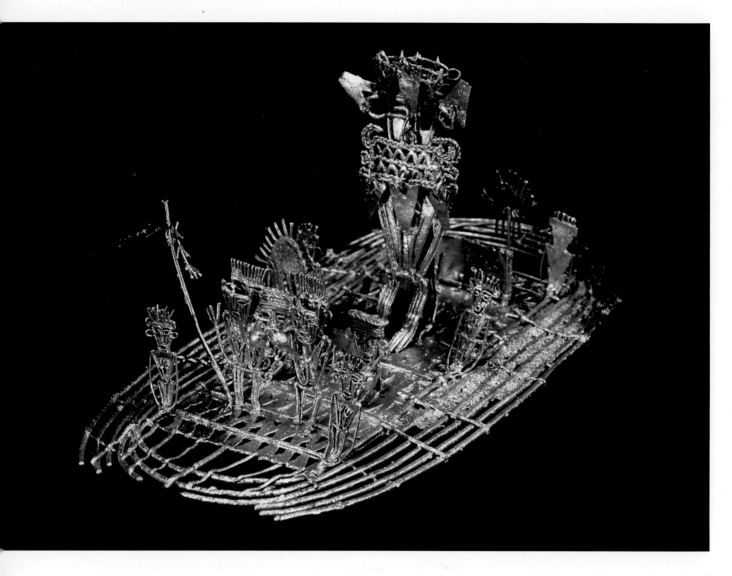

We have discovered a number of decorative objects made of gold, all very beautiful, particularly in the last two months, as well as an impressive collection of emeralds, chains, collars, and very unusual pieces of pottery.[14]

Years later, in Pasca, in the Cundinamarca department in the eastern cordillera of Colombia, peasants discovered a ceramic altar of repose surrounded by various objects, among them an extraordinary votive figure: a gold raft, on which was a central character surrounded by an entourage. This object went some way to confirming the reports about the "Golden Indian." Since then, it has been among the most important objects in the Museo del Oro del Banco de la República and is one of the symbols of the Colombian nation.

Bibliography

Botero, Clara Isabel. *The Construction of the Pre-hispanic Past of Colombia: Collections, Museums and Early Archaeology, 1823–1941*, Ph.D. dissertation, Oxford University, 2001.

———. *El Redescubrimiento del pasado prehispánico de Colombia: Viajeros, arqueólogos y coleccionistas 1820–1945*. Bogotá: Instituto Colombiano de Antropología e Historia, Universidad de los Andes, 2006.

14. Ibid.

Africa and Africans in the Colonial Mining World

DAVIKEN STUDNICKI-GIZBERT
Professor, World History
Department of History, McGill University, Montreal

I N THE LATE MIDDLE Ages, Europe was poorly supplied with precious metals. There was modest silver mining in German and Central European lands, but gold overwhelmingly came from abroad. Gold coinage was alloyed with other metals – a form of devaluation caused by scarcity – and many kingdoms simply lacked enough gold to mint their own.[1] In the Islamic Mediterranean, the situation was quite to the contrary thanks to the ready access that Muslim traders enjoyed to the mines of East Africa, the Sudan, Ghana, and Mali.[2]

The distance that lay between Europe and sub-Saharan Africa also encouraged the development of representations that would strongly prefigure the American myth of El Dorado. In an Italian text of the early fifteenth century, the royal hall of Prester John, Europe's fabled Christian ally in Ethiopia, was described as built of massive gold.[3] Like many of the mythical representations of gold that would appear in the Americas, these representations were supported by an underlying bedrock of fact. The gold fields of Africa were extensive and well exploited by numerous states and kingdoms, and this basic information did circulate, thanks again to the reports and texts of Muslim traders and travellers, through European consciousness. Perhaps the most famous case is that of Mansa

Musa, the king of the gold-rich kingdom of Mali. A devoted Muslim, Musa made hajj to Mecca in 1324–25 at the head of a long train of retainers and camels, over one hundred of which were reportedly loaded with gold. This discharge of Malian gold in the cities of Egypt and Arabia was sufficiently large to cause a surge of inflation throughout the Islamic Mediterranean.[4] The episode was duly noted by Italian merchants in Alexandria and was famously commemorated in Abraham Cresques's *Atlas* of 1375. The lower scene of the *Atlas* is dominated by the depiction of Musa, his gold throne, sceptre, and crown, and the arcing path of gold stretching from Ghana to Egypt, elements that all powerfully evoked the astounding wealth of faraway Africa.[5]

It was in the fifteenth century that the Portuguese finally succeeded in establishing direct contact with sub-Saharan gold. The new connections had a number of important consequences. The first was the sudden and increasing influx of gold into the European economy, a flow of bullion that began to resolve the problem of gold scarcity and contribute to the continent's commercial expansion.[6] Significantly, the Portuguese Crown began to strike gold coinage – the *cruzado* – for the first time in 1457.[7] This was the context surrounding the early career of Columbus, the signal figure of Iberia's

1.   Pierre Vilar, *A History of Gold and Money, 1450 to 1920*, translated by Judith White (London: Verso, 1991), pp. 30–31.
2.   Maurice Lombard, "L'or musulman du Ve au XIe siècle," *Annales. E.S.C.* (1947), 145–60; Anne McDougall, "The View from Awadaghost: War, Trade, and Social Change in Southwestern Sahara from the Eighth to the Fifteenth Century," *Journal of African History*, 26 (1985), 1–26; Jean Devisse, "Trade and Trade Routes in West Africa," in *UNESCO General History of Africa*, vol. 3 (London, 1988).
3.   Richard C. Trexler, *The Journey of the Magi: Meanings in History of a Christian Story* (Princeton: Princeton University Press, 1997), p. 128.
4.   Erik Gilbert and Jonathan T. Reynolds, *Africa in World History: From Prehistory to the Present* (Upper Saddle River, NJ: Pearson-Prentice Hall, 2004), p. 92.
5.   Jean Michel Massing, "Observations and Beliefs: The World of the Catalan Atlas," in Jay A. Levenson (ed.), *Circa 1492: Art in the Age of Exploration* (New Haven: Yale University Press, 1991); Helen Wallis, "Cartographic Knowledge of the World in 1492," *Mariner's Mirror*, Vol. 78, No. 4 (1992), pp. 407–18.
6.   Janet L. Abu-Lughod, *Before European Hegemony: The World System A.D. 1250–1350* (Oxford: Oxford University Press, 1991).
7.   Carlo M. Cipolla, *Before the Industrial Revolution: European Society and Economy, 1000–1700* (New York: W.W. Norton & Company, 1993), pp. 174–75.

49
Gold in the Americas

Slaves boarding a slave ship
The spouses of the slaves are in despair as their loved ones are taken away.

Photograph: Leemage

expansion across the Atlantic. In 1481, he made the first of a series of voyages to São João de Minas (Saint John of the Mines) on the Guinea coast. The trade that he was involved in had a number of pieces – alcohol, textiles, slaves, and spices – but the most important component by value was gold.[8] Columbus's voyage to the Americas can be read as an attempt to reproduce Portugal's African experience through a new westward route. His experience in the West African trade, combined with the swirl of ideas, images, and myths that surrounded African gold, deeply conditioned his perceptions and actions in the Caribbean.

In the Americas, the single most important group in the colonial gold mining industry was composed of the different nations of Africans brought over as slaves from West Africa. In the two principal gold mining regions, Nueva Granada and Brazil, they comprised the bulk of the labour force. In other regions, such as Mexico, Central America, Peru, and Chile, Africans and people of African descent also played central roles as foremen and bosses of the *cuadrillas*.

The prevalence of Africans in gold mining is in part explained by the demographic decline of local indigenous populations. This was clearly the situation in Minas Gerais, Brazil, where Africans were brought in as a substitute labour force, much to the detriment of the colony's established sugar- and tobacco-producing regions.[9] Nevertheless, the recruitment of Africans was due also to the skill sets and experiences that they had acquired in the gold-

---

8. Samuel Eliot Morison, *Admiral of the Ocean Sea: A Life of Christopher Columbus* (Boston: Litttle, Brown and Company, 1942), pp. 35–42.
9. Kathleen J. Higgins, *"Licentious Liberty" in a Brazilian Gold-Mining Region: Slavery, Gender, and Social Control in Eighteenth-Century Sabará, Minas* Gerais (University Park: Pennsylvania State University Press, 1999), pp. 30, 33.

mining districts of Africa itself.[10] These skills were part of the art of prospecting, which revolved around the capacity to "read" the landscape – evaluation of the subtle shifts in topography, the different gradations in the characteristics of the subsoil (often through smell), and the revealing changes in plant cover.[11] Even the material culture of placer gold mining, with its specialized tools – the *barreton*, an iron spatulate blade, and the *almocafre*, a hand-held claw-like cupping tool – was heavily inflected with African influences.[12] In Minas Gerais, female slaves from the Gold Coast were sought out because of their "magical" ability to detect rich ore beds that lay beneath the gravel and mud. "For this reason," wrote the governor of Rio de Janeiro in July 1726, "there is not a Mineiro who can live without a Negress from Mina, saying that only with them do they have any luck."[13] The high regard with which these women were held in the gold fields of the Americas has interesting transatlantic resonances. In the Mande-speaking societies of eighteenth-century Guinea, Mali, and Sierra Leone, the working of placer mines was undertaken exclusively by women, and gold in general was associated with femininity, fertility, and the woman's sphere.[14]

10   Kris Lane, *Quito 1599. City and Colony in Transition* (Albuquerque: University of New Mexico Press, 2002), pp. 74, 123.

11   Robert C. West, *Colonial Placer Mining in Colombia* (Baton Rouge: Louisiana State University Press, 1952).

12   Personal communication, Kris Lane, Department of History, College of William and Mary.

13   Charles Boxer, *The Golden Age of Brazil, 1695–1750: Growing Pains of a Colonial Society* (Berkeley: University of California Press, 1964), p. 165.

14   B. Marie Perinbam, "The Salt-Gold Alchemy in the Eighteenth and Nineteenth Century Mande World: If Men Are Its Salt, Women Are Its Gold," *History in Africa,* Vol. 23 (1996), pp. 259, 265.

Detail of the marine map of the Atlantic Ocean
by Pierre de Vaux, 1613

Photograph: AKG Paris

Mukis:
Masters of the
Andean Mines

DAVIKEN STUDNICKI-GIZBERT

Professor, World History
Department of History, McGill University, Montreal

"WHO ARE YOU?" I asked.
And a rough and powerful voice answered:

"I am Tayta Muki, master of the ore bed."

It was then that I saw a very small man whose body glowed like gold, with two horns peeping out from his hat and burning red eyes.

"Do you desire wealth and metal?" Tayta Muki asked. "It's easy. I can show you the veins. I can work it for you and leave it for you all ready. I only ask for a life of a man within the next two years and that when you visit you bring me my pleasures: coca, cigarettes and alcohol."

This is the account given by a contemporary Bolivian miner to the French anthropologist Carmen Salazar Solar regarding the *muki*, or masters of the Andean mines. These were gnomic beings the size of a child, of golden complexion, brilliant red eyes and a disproportionately long penis that either was wrapped, belt-like, around their waist or coursed, for kilometres, along the veins of precious metals. They were responsible for fertilizing the veins and caring for Pacha Mama, or Earth Mother. As protectors of the mines, the *muki* were powerful figures. They could cause cave-ins or move ore-bearing rocks hundreds of kilometres away through a network of tunnels that connected the different mining centres of the Andes. Or, alternatively, they acted as patrons to miners, leading them to rich veins or warning them of impending danger. Unsurprisingly, miners took particular care when entering the mines to properly appease the *muki* through ceremonies and regular offerings to effigies made in their image.[1]

The figure of the *muki* continues to be present in the contemporary mining cultures of the Andes.[2] Reconfigured by the passage of the centuries, he is now known as *Tio*, or Uncle, and his persona has been mixed with Christian figures such as the devil and, more recently, an iconic *gringo* mine-owner, engineer, or administrator. Effigies of the *Tio*, sculpted of ore with eyes made of light bulbs or shiny-bright tin, can be found at the mouths of mine shafts across Bolivia, waiting to be propitiated with offerings of alcohol, cigarettes, and coca.

◄ Bolivian farmer making an offering to Muki
the god who protects and miners

Photograph: ©Loren McIntyre

1.   B. Marie Perinbam, "The Salt-Gold Alchemy in the Eighteenth and Nineteenth Century Mande World: If Men Are Its Salt, Women Are Its Gold," *History in Africa*, Vol. 23 (1996), pp. 239–51, 325.

2.   See June Nash, *We Eat the Mines and the Mines Eat Us: Dependency and Exploitation in Bolivian Tin Mines* (New York: Columbia University Press, 1979); Michael T. Taussig, *The Devil and Commodity Fetishism in South America* (Chapel Hill: University of North Carolina Press, 1980).

CHAPTER **3**

THE CONQUERORS

Entra CORTES con su exercito en S. Mexico, y es recibido por
Motezuma con muestras de grande amistad.

The Conquistadors

Paz Cabello-Carro*
Director, Museo de América de Madrid
* With the collaboration of Hélène Dionne

Translated by Käthe Roth

The Conquistadors

THE CONQUISTADORS WERE MEN of arms, explorers, and adventurers who seized immense territories in the New World in the name of the Spanish Crown. A good number of them were *hidalgos* (noblemen with little wealth), and most came from the Extremadura region. Even though, in the Western imagination, the word "conquistador" is applied indiscriminately to all Spanish soldiers and adventurers, only the leaders of expeditions had the right to this title.[1]

On every expedition, there was a literate person who was responsible for recording and describing events. A number of these accounts have survived to the present. The expeditions of exploration, conquest, and colonization were mainly private undertakings. Expedition leaders had to raise the funds, obtain the royal permissions necessary, and assume all the risks. Recruitment was often based on family and neighbourly relations. Once the territory was conquered, the settlers invited to move there were also chosen from family and the region, their common roots making it easier for them to settle in America.

> Because of their obsession with gold, the conquistadors are often dismissed as "gold crazy." In fact they were not so much gold crazy as status crazy. . . . To obtain these royal favors, their expeditions had to bring something back to the king. Given the difficulty and expense of transportation, precious metals

◄ Cortés enters Mexico City with great pomp and is warmly greeted by Motecuhzoma

Oil on copper
1776–1800
Museo de América, Madrid, Inv. 00209
Photograph: Museo de América

– "nonperishable, divisible, and compact," as historian Matthew Restall notes – were almost the only goods that they could plausibly ship to Europe. Inka gold and silver thus represented to the Spaniards the intoxicating prospect of social betterment.[2]

Among all the Spanish military leaders, two conquistadors have been the most prominent in the Western imagination. They are Hernán Cortés, conqueror of the Aztec kingdom, and Francisco Pizarro, conqueror of the Inca empire. Here are brief descriptions of their exploits.

Hernán Cortés

Fernando Cortés Monroy Pizarro Altamirano, or Hernán Cortés, was born in Medellin, Spain, in 1485. As a teenager, he heard about the exploits of Columbus and other navigators and conquistadors who returned from the East Indies with fabulous stories. The son of a *hidalgo*, he could not, as a noble, choose a manual trade. He was therefore sent to Salamanca to study the law. After two years, he abandoned his education and went to Seville; there, he signed on with Nicolás de Ovando, who was preparing for an expedition to explore the New World.

In 1504, Cortés was in Hispaniola. Don Diego de Velasquez, conquistador and governor of the island, invited him to join his next expedition to Haiti and Cuba. For a number of years, under the command of Velasquez, Cortés navigated in the Caribbean, waging war and subduing the indigenous populations. Velasquez then decided to organize expeditions to the north and west on the mainland. Eventually, he gave Cortés command of an army to go to free Grijalva, a compatriot who had fallen into the hands of hostile Indians. In addition to this mission, Cortés was to make

---

1.   "Conquistador" is neither a military rank nor a title of nobility, but a symbolic figure, used widely in literature, historical narration, cultural criticism, and Western fiction.
2.   Charles C. Mann, *1491: New Revelations of the Americas Before Columbus* (New York: Knopf, 2005). p. 81.

The Spanish conquistador Hernán Cortés (1485–1547)

Anonymous, late sixteenth century
Photograph: The Art Archive/Academia BB AA S Fernando Madrid/
Gianni Dagli Orti

contact with the Indians, subdue them, and impose on them a tax in gold, pearls, and gems. Sailing along the Yucatan coast, Cortés landed at Veracruz in the spring of 1519. After a battle in which Spanish arms and horses overwhelmed the Totonacs, he negotiated with them and was offered gold, livestock, and slaves. One of the slaves was an Indian woman named Malintzin (Malinche), who became his interpreter, advisor, and lover. Continuing his mission of exploration, he heard talk in Tabasco of a kingdom farther west, called Mexico-Tenochtitlán, which was overflowing with gold. He continued along his way, leaving in his wake the terrible memory of bloody massacres. At Tlaxcala, he formed an alliance with the Tlaxcaltecs, who were enemies of the Aztecs.

The emissaries of the Aztec emperor, Motecuhzoma II Xocoyotzin, went to meet with Cortés and talked to him about Mexico-Tenochtitlán. At their second meeting, he managed to arrange an introduction to the emperor. He then understood

that the key to power in Mexico was before him and that he had to try to seize the land and immense riches of the kingdom of Motecuhzoma II. Cortés entered Mexico-Tenochtitlán with great pomp in November 1519. This city was the largest that the Spaniards had ever seen: "And yet the capital of the Aztecs, Tenochtitlán, was five times larger than Madrid and twice as populous as Seville, the largest city in Spain."[3] Enthralled by the wealth that he saw, Cortés demanded more gold. He and his captains moved into the palace, where they found a bricked-up room. It was the former emperor's treasure room. Thrilled by the fabulous riches, Cortés was nevertheless aware that the Aztecs might assassinate him. Devising a plot with his men, he took the emperor hostage, forcing him to swear allegiance to Charles Quint and to pay a tax each year as a vassal. In doing this, Cortés had overstepped his bounds. He did not have royal permission to conquer and submit the populations that his reconnaissance mission gave him the right to encounter. Dissent arose among his men, and among his compatriots who had stayed in Cuba.

When Cortés had to return to Veracruz to intercept and stop a new expedition that had arrived from Cuba, the Indians rebelled following the massacre at the Toxcatl festival and demanded that their emperor be freed. In a heroic charge, Cortés and the troops gathered at Veracruz broke through the blockade of Mexico City to reach the Spaniards besieged there. To negotiate his return to the coast, Cortés ordered Motecuhzoma II to speak to his people. Caught in a trap, Cortés and his men made a desperate attempt to escape. However, they did not want to leave behind their marvellous booty, and they tried to push through thousands of Aztecs, their horses loaded with gold and weapons. The confrontation was brutal. More than 2,000 allied Indians and 600 Spaniards were killed or captured and sacrificed on the terrible night of June 30 to July 1, 1520, the notorious "Noche Triste." Cortés managed to flee with a handful of men, abandoning his dream and his riches. Nevertheless, he did not consider himself defeated. He rallied to his cause all the enemies of the Aztecs and took his revenge with an orchestrated attack, assaulting Mexico-Tenochtitlán from all sides. At the end of a three-month siege, the last Aztec emperor, Cuauhtémoc, surrendered on August 13, 1521.

---

3.   Eduardo Galeano, *Les veines ouvertes de l'Amérique latine. Une contre histoire* (Paris: Plon, 1981), p. 28.

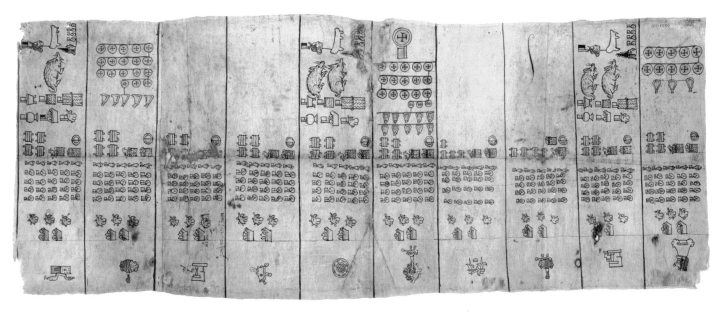

After making him master of the Aztec empire, the king of Spain gave Cortés the title of master of the estate of Marquesado del Valle. Cortés then undertook a number of expeditions in search of a passage linking the Atlantic to the Pacific. His explorations took him to the Gulf of California, or Sea of Cortés. He died on December 2, 1547, at Castilleja de la Cuesta, Spain, where he was organizing a new expedition.

Francisco Pizarro

Francisco Pizarro González, Marqués de los Atabillos, was born in Trujillo, Spain, in 1476. An illegitimate son of the navigator Gonzalo Pizarro Rodriguez de Aguilar, a member of the Spanish low nobility, he was a member of a large family linked to that of Cortés, of whom he was a distant cousin. Raised as a peasant rather than a noble, Francisco Pizarro did not have the privilege of an education; although he was very intelligent, he remained illiterate all his life. Like his father, he chose a military career, and both were engaged in the campaign in Italy. He hoped that his exploits would earn him titles, responsibilities, and pensions from the Crown of Spain.[4]

In 1502, Pizarro left for America with Nicolás de Ovando. He was clever and rapidly rose through the ranks. He took part in major expeditions: after travelling with Alonso de Ojeda to the Gulf of Uraba, he joined Vasco Nuñez de Balboa, who discovered the Pacific in 1513. After a few years of campaigns of all sorts, he obtained an *encomienda* of Indians and moved to Panama, where he made a good profit raising livestock.

After several years, however, Pizarro became bored with the life of the landowner. He missed action. The account describing the fabulous riches that Cortés had seized during the conquest of Mexico was circulating among the Spaniards living in the Caribbean, raising wild hopes. Pizarro re-enlisted as a captain and invested his fortune in organizing expeditions to find the "country of gold." In 1524, he joined forces with Diego de Almagro, another conquistador, to lead the undertaking, and a priest, Hernando de Luque, a clerk in the Panama diocese, who supplied the funds needed for these enterprises, as well as the governor's authorization. Efforts, energy, and endurance were needed both to maintain the troops' morale and to overcome obstacles such as climate, mountains, forests, and hostilities with the Indians. In 1526, Pizarro arrived in the city of Túmbez and finally saw his efforts rewarded. The wealth of the Incas was within reach. But, to conquer this kingdom

---

4. Mann, 2005.

Portrait of Francisco Pizarro

Anonymous, sixteenth century

Museo de América, Madrid, Inv. 00185

H: 44 cm W: 39 cm

Photograph: Museo de América

and appropriate its treasures, he had to obtain *capitulaciones,* or royal orders. In 1529, therefore, Pizarro left for Spain, where he obtained an audience with Charles Quint.

In 1530, Pizarro returned to Panama accompanied by his brothers, Hernando, Gonzalo, and Juan Pizarro, and his half-brother, Francisco Martin de Alcántara. In late January 1531, three armed ships transporting 180 men and 37 horses set sail for Peru. Pizarro and his men landed in the middle of a conflict of imperial succession.

Two of the sons of the great Inca emperor Huayna Capac, Huascar Capac and Atahualpa, were fighting over the imperial throne. After bloody battles, Huascar, sensing that victory was slipping away, decided to flee Qosqo (Cuzco) with his army. Atahualpa sent his troops to intercept him

and take him prisoner. Huascar was brought back to Cuzco and saw his family executed before his eyes. Victorious, Atahualpa decided to return with great pomp to his northern palace, in Ecuador. On the way, he stopped over in the small town of Cajamarca, where he learned that pale, hairy men had landed on the coast and were mounted on enormous beasts. Atahualpa was curious, and he decided to wait for these strangers to arrive. Pizarro sent emissaries and asked to meet the future emperor. When Atahualpa arrived in the central square of Cajamarca in his golden litter, surrounded by 5,000 men in ceremonial costumes and arms, he had no idea that Pizarro would ambush him with his horsemen and cannons. A priest in the Spanish party handed the Inca a breviary soiled by the vagaries of the voyage. Atahualpa, a pure Indian, was offended by this lack of respect and threw the sacred book away. This was the excuse that the Spanish needed to cry sacrilege and open fire. In the panic and confusion, Atahualpa was thrown out of his litter and captured by Pizarro.

After lengthy negotiations, Atahualpa realized that the Spanish were obsessed with gold and offered to trade his freedom for a fabulous "ransom." He told his men to go and fetch the treasure sitting in the palace in Cuzco. The territory was vast and it took many weeks before the gold, so eagerly awaited, was brought to Cajamarca. The treasure handed over to the Spaniards surpassed all their hopes.

"The conquistadors were also great experts at the techniques of treason and intrigue. They knew . . . how to exploit to their advantage the division in the Inca empire between Huascar and Atahualpa, the warring brothers. They depended on accomplices among the dominant intermediary classes, priests, functionaries, military men, once the murder had disorganized the indigenous dignitaries."[5] Fearing that Atahualpa, once freed, would turn against them, Pizarro had him executed after an expedited trial and hurriedly placed an allied prince on the throne. On August 29, 1533, with the death of Atahualpa, the Inca empire was vanquished.

How had 138 conquistadors and three dozen horses been able to crush such a great empire and put an end to a magnificent civilization? Had it not been for the horses and cannons, Pizarro, even with his taste for intrigue and all-consuming ambition, would not have beaten the much larger Inca army.

---

5.   Galeano, 1981: 29.

ATAHUALLPA, INCA XIIII.

The fact that the empire had been ravaged by inter-necine wars over succession to the imperial throne had set up a chessboard on which Pizarro and his men, profiting from the political controversy, simply had to make their moves to take control of the battle and checkmate the opponents.

On the way back, Pizarro founded Ciudad de los Reyes (the city of kings), today Lima, on January 5, 1535. His brother Hernando returned to Spain to report the exploits of the conquest of Peru to the king and give him his share of the fabulous treasure. The king granted Pizarro the title of marquis and offered him the governorship of Peru. His reign as governor was not without problems. Many revolts broke out, and Pizarro was assassinated on June 26, 1541.

Portrait of Atahualpa, thirteenth and last king of the Incas

Anonymous, sixteenth century
Ethnologisches Museum, Staatliche Museen zu Berlin, Inv. V A 66707
Photograph: Dietrich Graf, Bildarchiv Preussischer Kulturbesitz/
Art Resource, New York

Circulus articus:

Tera del Rey de portugall

Mare germanieo

del Rey de castella:

Este he o mar o ẽtre castella, e portugall

dao del Rey de castella

A linha equinocialis:

Mare occanus:

ycone.

Pollus antarticus:

Fray Bartolomé de las Casas (1484–1566), or the Debate Surrounding the Legitimacy of the Spanish Conquest

MIGUEL LUQUE TALAVÁN

Professor, Department of American History, Universidad Complutense de Madrid

Translated by Käthe Roth, from the French translation by Louis Jolicœur

A LL SCHOLARSHIP ON THE subject of Fray Bartolomé de las Casas highlights themes related to the conquest and colonization of America and, in particular, the controversy surrounding the legitimacy of the conquest itself.

In the sixteenth century, the Crown of Castile brought together the most important scholastic authors since St. Thomas Aquinas.[1] If one had to set a historical moment marking the inception of legal literature around the question of the "just war," one would no doubt choose the famous reaction of Queen Isabelle of Castile when the first Indians from America arrived in Spain in April 1495: "Were they captured in a just war?"[2]

From that time until 1573, jurists and theologians steered the medieval debate on the situation created by the discovery of America, giving their theories a clearly moralizing content. One must remember that Portugal's and Spain's discoveries

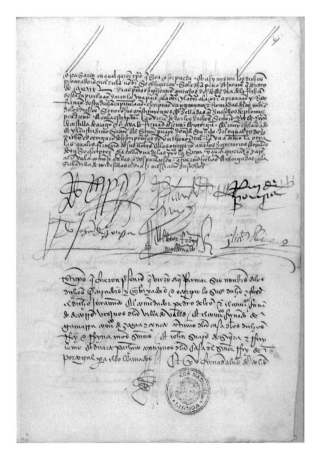

◄ Map made by Sebastiano Cantino in 1502

showing the line of demarcation imposed by the Treaty of Tordesillas (1494) between Portuguese America and Spanish America. The line of demarcation was a north-south meridian located 370 leagues (1,770 km) west of the Cape Verde Islands, today situated at 46° 37' west. This is why Brazil became Portuguese.

Biblioteca Estense, Modena

Photograph: © Costa/Leemage

The international Treaty of Tordesillas,

signed June 7, 1494, dividing the New World between Spain and Portugal, was written under the aegis of Pope Alexander VI.

Photograph: Lisbon National Library.

---

1. Lewis Hanke, *Cuerpo de documentos del siglo XVI. Sobre los derechos de España en las Indias y las Filipinas*, discovered and annotated by Lewis Hanke, collected by Agustín Millares Carlo (Mexico City: Fondo de Cultura Económica, 1977), p. xiv.

2. Ibid., p. xv. Antonio Muro, "Normas de justicia en las guerras contra los indios," *Actas del I Simposio sobre "La ética en la conquista de América (1492–1573)" (Salamanca, 2-5 nov. 1983)* (Salamanca: Excmo. Ayuntamiento y Excma. Diputación Provincial de Salamanca, 1984), p. 550. Joaquín F. Pacheco, Francisco De Cárdenas, and Luis Torres de Mendoza (eds.), *Colección de documentos inéditos relativos al descubrimiento, conquista y colonización de las posesiones españolas en América y Oceanía, sacados, en su mayor parte, del Real Archivo de Indias bajo la dirección de los Sres. D. Joaquín F. Pacheco y D. Francisco de Cárdenas ... y D. Luis Torres de Mendoza* (Madrid: Imprenta de Manuel Bernaldo de Quirós-Manuel G. Hernández, 1864–89), vol. 38, p. 342.

were conducted under standards derived from Roman and canonical law. The Spanish Crown had three titles of sovereignty: discovery and taking of possession (1492), pontifical and alexandrine bulls (1493), and the Treaty of Tordesillas (1494).

These titles were the subject of very intense controversy, as borne out in the legal literature. On one side were authors who supported the titles granted by the Crown. On the other side were mainly theologians, such as Fray Francisco de Vitoria and las Casas, who criticized not only the titles, but, by this very fact, the right of Spain to colonize America. The Spanish theologians of the time referred to theses of medieval scholasticism, in particular St. Thomas Aquinas, to define principles for the conduct of Spaniards in their contacts with the inhabitants of the Americas. According to these authors, indigenous peoples should not be dispossessed of their authority or their goods, and even the alexandrine bulls were open to question. Thus, it was the theologians who first criticized the first two titles of sovereignty in the West Indies. In the case of the third title, the critique came mainly from the French, English, and Dutch, for whom the Treaty of Tordesillas had no value. The scholastic theologians proposed other titles of sovereignty, but without success, so the Crown maintained its initial titles throughout the Spanish presence in the Americas. The controversy was settled in 1573: in the ordinances of that year, the word "conquest" was replaced by "pacification." This measure was to give a new direction not only to Spanish penetration into the West Indies, but also to how authors of legal literature discussed the theme.

At the time, Spain was the only nation in which there was debate over the legality of the process of conquest and colonization. This has been interpreted from various angles at different times and in different historiographic currents. Some see works by authors such as Vitoria as an attempt to ". . . counter the black legend of Spanish domination in America, to deny the decadence of scholasticism in the sixteenth century, even to highlight the qualities of the Hispanic imperial era."[3] Other

Morion cabasset
Italy, 1570–90, Inv. 12
H: 31.7 cm L: 36.2 cm W: 25.4 cm
Photograph: Courtesy of Higgins Armory Museum, Worcester, Massachusetts

historians, however, are of the opinion that these debates were encouraged by the Crown simply to find legal and theological foundations for the actions of the Spanish monarchy in the Indies.[4]

In this debate, a few authors, such as Luis Vives, Vitoria, las Casas, and Fray Domingo de Soto, stated that all wars are illegal, inhuman, and absurd. Others, notably Alfonso Álvarez Guerrero, Fray Juan Ginés de Sepúlveda, and Bernardino de Arévalo, believed that war could be just if certain conditions are met. Finally, Diego de Covarrubias y Leyva, without addressing directly the issues related to the conquest, maintained that individuals could not be legally reduced to slavery except through a "just war."[5]

Fray Bartolomé de las Casas was born in Seville in 1484, the son of an explorer who had accompanied Christopher Columbus on his second (1493)

---

3. José Miranda, *Vitoria y los intereses de la conquista de América* (Mexico City: El Colegio de México [Jornadas; 57], 1947), p. 5.
4. "The new developments in theology in Spain tended over all to justify or legitimize reality. The forced rehabilitation of the old concepts was aimed at adapting them to the needs of these theoretical justifications or legitimizations" (ibid., p. 5, note 2).
5. Hanke, 1977, p. xv. For a brief analysis of the main Spanish works dealing with the "just war," see ibid., p. 17–46. Venancio Diego Carro, *La Teología y los teólogos-juristas españoles ante la conquista de América* (Madrid: Tall. Marsiega, 1944).

La Brevísima relación de la destrucción de las Indias

Title page of a book by las Casas

Bibliothèque et archives nationales du Québec,
Direction générale de la conservation
Photograph: Bibliothèque et Archives nationales du Québec

and third (1500) voyages to America. After studying grammar in the town where he was born, he went to Santo Domingo in 1502. As a reward for having done battle against the indigenous peoples, he obtained an *encomienda*.[6] A few years later, in 1506, he accompanied Bartholomew Columbus, the admiral's brother, on his voyage to the Vatican. In Seville, he received minor orders; he was ordained in Rome in 1507. He began to celebrate the mass only in 1510, however, once he returned to Hispaniola. In 1512, he went to Cuba to take the position of military chaplain, while preserving his *encomienda*. He began to defend the rights of indigenous Americans in 1514. After renouncing his *encomienda*, he travelled to Spain with Fray Antonio de Montesinos O.P.[7]

Even before the question of "just titles" arose, the methods used to occupy and conquer the Indies were raising controversy.[8] The debate began on the fourth Sunday of Advent, 1511, with the famous sermon given by Father Montesinos before Diego Columbus and all of the other representatives of the Spanish authority on Hispaniola, but supported by his entire community. "I am the voice of Christ in the desert of this island," he began; he then called upon the faithful to recognize the rational souls of the indigenous peoples and asked them by what right they made war against them and forced them to work. The following Sunday, he gave another sermon in the same vein.[9]

The controversy over the legitimacy of the Spanish action generated a number of legal opinions dictated by the Crown and by various subordinate institutions, including provincial and diocesan councils, with the goal of regulating

6. Translator's note: The institution of *encomienda* consisted of dividing the Indians into groups under the command of an *encomendero*. The Indians had to pay a tax to and work for the *encomendero*, who was charged with protecting and evangelizing them.

7. José Alcina Franch, *Bartolomé de las Casas* (Madrid: Historia 16: ed. Quórum: Sociedad Estatal para la Ejecución Programas del Quinto Centenario, 1986).

8. Zavala and Hanke offer a detailed analysis of the evolution of the "just war" doctrine and its application in the context of the Spanish conquest. See Lewis Hanke, *La lucha española por la justicia en la conquista de América* (Madrid: Aguilar, 1967), pp. 230–51; Silvio A[rturo] Zavala, *Las instituciones jurídicas en la conquista de América.* (2nd ed., rev. and exp.) (Mexico City, D.F.: ed. Porrúa [Biblioteca Porrúa; 50], 1971), pp. 435–86. It is interesting to note that Juan de Solórzano Pereyra, in his book *Política Indiana*, was the first author to deal with "just titles," in particular in chapters 9 to 12 of vol. 1, *Política Indiana compuesta por el Señor Don Juan de Solórzano y Pereyra, Cavallero del Orden de Santiago, del Consejo de su Majestad en los Supremos de Castilla e Indias Corregida, é Ilustrada con Notas por el Licenciado Don Francisco Ramiro de Valenzuela, Relator del Supremo Consejo, y Cámara de Indias, y Oidor Honorario de la Real Audiencia, y Casa de la Contratación de Cadiz* (Madrid, 1736); Estudio Preliminar por Miguel Ángel Ochoa Brun (Madrid: Biblioteca de Autores Españoles desde la Formación del Leguaje hasta nuestros días, vol. 252), 1972.

9. Francisco Morales Padrón, *Teoría y leyes de la conquista* (Madrid: ed. Cultura Hispánica del Centro Iberoamericano de Cooperación, 1979), pp. 306–07.

Silver stirrups

1695–1800, Musée des beaux-arts de Montréal, Inv. 1962.Ds.14 and 1962.Ds.15
L: 24.9 cm W: 10.8 cm
Photograph: Christine Guest
Musée des beaux-arts de Montréal

the discovery, occupation, and settlement of territories still unknown or unconquered.[10] Las Casas was one of the most important witnesses to these events.[11] No doubt because of the great moral influence of Father Montesinos, and the Dominicans in general, Father las Casas entered the Dominican order, to which the greatest jurists of the school of Salamanca belonged, in September 1522. From then on, both in his engaged life and in his major works, characterized by a scholastic and humanist style, las Casas was to devote himself to a single cause, protection of the rights of indigenous peoples,[12] even though he participated in what Beuchot calls "social experiments with colonization," such as the settlement at Cumaná, which was a failure, and pacification of the indigenous people at Vera Paz.

Las Casas refused to be named bishop of Cuzco (Peru); he was approached to become bishop of Chiapas, but turned down this position in 1550. From his diocesan seat, las Casas frequently confronted the region's *encomenderos*, who refused to apply the "new laws" (1542–43) that greatly interfered with their interests but improved the situation of the indigenous peoples. It was during this period that he wrote one of his best-known works: *Brevísima relación de la destrucción de Las Indias*. First published in Seville in 1552, the book was widely distributed in Europe. It was translated into Dutch, French, Latin, German, English, and Italian and was used as a political weapon by all nations that questioned the Spanish presence in America.

In 1547, las Casas returned permanently to Spain, where he continued to write tirelessly in favour of the indigenous peoples. This courageous position contrasted with his proposal to have African slaves brought to America (a proposal founded on Aristotelian and Thomist doctrines) to better protect the indigenous peoples. In the end, he rejected this option when he denounced to King Philip II the violations committed against African slaves.

Unlike other theorists of his time, las Casas maintained that the indigenous peoples "possess a high level of rationality, as evidenced by their great cultural wealth, which clearly indicates that they are capable of governing themselves."[13] This was the basis of his famous defence of his position, against

10. Alberto de la Hera [Pérez-Cuesta], "La 'Guerra Justa' y la polémica sobre los métodos," in Ismael Sánchez Bella, Alberto de la Hera [Pérez-Cuesta], and Carlos Díaz Rementería, *Historia del Derecho Indiano* (Madrid: Fundación MAPFRE América, ed. MAPFRE [Colecciones MAPFRE 1492], 1992), pp. 145–46.

11. Bartolomé de las Casas, *Brevísima relación de la destrucción de Las Indias* (Ed. d'André Saint-Lu. Madrid: Cátedra [Letras Hispánicas], 1989).

12. For a brief analysis of las Casas's work, see Francisco Esteve Barba, *Historiografía indiana* (2nd ed.) (Madrid: ed. Gredos, 1992), pp. 83–102.

13. Mauricio Beuchot, *La querella de la conquista. Una polémica del siglo XVI* (Mexico City: D.F.: Siglo Veintiuno ed. [América nuestra; 38], 2004), p. 59.

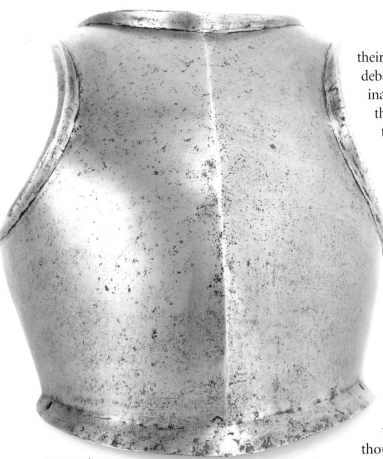

Armoured vest

Probably Italian, 1500–10, Inv. 2002.01.1

H: 41.9 cm L: 34.3 cm D: 12.7 cm

Photograph: Courtesy of Higgins Armory Museum,
Worcester, Massachusetts

their descendants, one of the most controversial and debated aspects at the beginning of Spanish domination.[15] In this regard, las Casas maintained that the indigenous nobles were "of rank as royal as the princes of Castile."[16] In spite of the number of arguments proposed against these rights, it is remarkable that the opinions expressed by Fray Bartolomé de las Casas, as well as by many authors throughout the sixteenth century, particularly the Franciscans, finally won the day. The Crown recognized the rights of indigenous peoples in 1557. It must be conceded, however, that a desire for justice was not the only factor in this decision. Political and economic motives also played a role in the royal decision in favour of "natural subjects":[17] the Crown had to count on the cooperation of indigenous peoples within the government of the Republic of Indians. Finally, although las Casas's positions were in line with those of an entire current of thought in favour of the indigenous world, he was always identified mainly with defence of the rights of the indigenous populations in the face of a social class, that of the conquistadors-*encomenderos*, that approved of their exploitation.[18]

Las Casas died in Madrid on July 18, 1566. His life and work fit within a context of conquest of the American space in which three, often conflicting, realities coexisted: the interests of the Crown, those of the Church, and those of the conquistadors-*encomenderos*.[19] Beyond the importance of the Salamanca school in general and of Fray Bartolomé de las Casas in particular, the work done at the time in favour of human rights endures to this day, since many indigenous populations in Latin America are still deprived of the most fundamental rights.

the contrary arguments of Sepúlveda, at the Junte de Valladolid of 1550 convened by Charles Quint to discuss the legitimacy of the conquest. Las Casas was firm in his vision of the human being, while Sepúlveda stated that the indigenous peoples were barbarians and that it was legitimate to make war against them. In his statement, las Casas referred to Vitoria and Soto, placing him among the disciples of the Salamanca school.[14]

It is also interesting to bring up las Casas's argument regarding recognition of "natural subjects" and

14. Ibid., pp. 59–68. Similarly, this debate must be linked to the perceptions that Europeans in the sixteenth century had of the American reality and the question of integrating the New World into the "mental horizon of Europe," as Sir John Elliott expressed it. In this regard, see Sir John H. Elliott, *El viejo mundo y el nuevo (1492–1650)* (Madrid: Alianza ed., 1990), "2. El proceso de asimilación," pp. 41–70; Sir John H. Elliott, *Imperios del mundo atlántico. España y Gran Bretaña en América (1492–1830)* (Madrid: Taurus, 2006), "Capítulo 3. Frente a los pueblos americanos," pp. 103–46.

15. On the issue of the "natural subject," see the excellent analysis by Carlos J. Díaz Rementería: *El cacique en el Virreinato del Perú. Estudio histórico-jurídico* (Seville: Publicaciones del Seminario de Antropología Americana, Departamento de Antropología y Etnología de América, Facultad de Filosofía y Letras, Universidad de Sevilla, 1977), "La problemática del señorío natural," pp. 53–57.

16. Antonio Dougnac Rodríguez, *Manual de Historia del Derecho Indiano* (Mexico City, D.F.: Universidad Nacional Autónoma de México, Instituto de Investigaciones Jurídicas [Serie C: Estudios históricos; 47], 1994), p. 325; "Carta de las Casas a Miranda," in A. M. Fabié, *Vida y escritos de Don Fray Bartolomé de las Casas* (Madrid, 1879), vol. 2, p. 602; Juan Belda Plans, *La Escuela de Salamanca y la renovación de la teología en el siglo XVI* (Madrid: Biblioteca de Autores Cristianos [BAC maior; 63], 2000); Luciano Pereña, *La Escuela de Salamanca* (Salamanca: ed. de la Caja de Ahorros y Monte de Piedad de Salamanca [Ensayo de Filosofía Política], 1986).

17. Delfina Esmeralda López Sarrelangue, *La nobleza indígena de Patzcuaró en la época virreinal* (Mexico: n.p., 1965), pp. 83–87.

18. Beuchot, 2004, p. 56.

19. On this question, aside from the work by José Miranda mentioned above, see Silvio A. Zavala, *Los intereses particulares de la Conquista de la Nueva España* (Mexico City, D.F.: El Colegio Nacional, 1991).

The Gold of the *Girona*

WINIFRED GLOVER
Curator of World Cultures,
The Ulster Museum, Belfast

IN 1588, IT SEEMED as if Spain controlled half the world. As well as its own conquests in the Americas, its conquest of Portugal and all its colonies in 1580 and control over large parts of Holland also gave it command over their trading stations.

England, however, had long been an irritation to the Spanish by constantly attacking its treasure galleons as they made their way home from the South American colonies laden with gold and silver bullion and rich ornaments made by the Native American goldsmiths. The Spanish king, Philip II, had tried to form an alliance with England by marrying the devout Catholic Queen Mary and, after her death, had even offered for the hand of the staunchly Protestant Elizabeth I. Her delaying tactics, coupled with the continued raids on the Spanish fleets and settlements by Sir Francis Drake and others, finally wore his patience down. The sober and devout Philip determined to invade England and to bring it under Spanish dominion.

When he embarked on the enterprise, he could not have foreseen the tragic outcome, with the noblest of his aristocracy being lost on the bleak rocks of a northern Irish shore. Philip unwittingly condemned the Armada to failure with his idealistic religious and political ambition, which showed little regard for military realism.

Even by modern military standards, the plan to crush England, not only with a large armada but also with a mighty army moving through Europe to rendezvous with the fleet in Flanders, would be extremely hazardous to execute. Inevitably, difficulties and delays between the naval and land commands were to seal the fate of the endeavour and present the advantage to the mobile defender.

The Armada

One hundred and thirty ships set sail from Lisbon in 1588 under the command of Don Alonso Perez de Guzman, Duke of Medina Sidonia. There were four enormous galleasses – ships intended to have the double advantage of being powered by sail and oar, but, in reality, difficult to manoeuvre – and small, fast sailing petaches, with a great variety of ship sizes in between. The ships all carried cannon as well as the weaponry necessary for an invasion and all the foodstuffs needed to sustain them. As well as sailors, soldiers, and armaments, they carried clerics, because Philip viewed the Armada very much as a religious undertaking – before it set sail, La Felicissima Armada was blessed and all on board received religious medals.

After a first engagement in the English Channel on July 31, followed by other attacks off Portland Bill

◀ **Bates Littlehales**
The *National Geographic* photographer spent several weeks with the Belgian underwater archaeological expedition in 1969.

Photograph: Marc Jasinski

▲ Spanish gold escudo bearing a cross potent, or Jerusalem cross, on its reverse
The symbolism of this cross, portraying the coat of arms of the Knights of the Hospital of St. John in Jerusalem, is that in all four corners of the world, each represented by a miniature cross, all derives from Unity and all returns to Unity.

Late sixteenth century
Photograph: Michael McKeown/The Ulster Museum

Naval historian and archaeologist Robert Stenuit

emerges from the water with one of the heavy gold chains found in the wreckage of the *Girona*. The chains were made of non-welded links and were used to make payments.

Photograph: Marc Jasinski

on August 2 and the Isle of Wight on August 3, the Armada's commander decided that the fleet should anchor at Calais and wait for news of the army under the command of the Duke of Parma. The English took their opportunity and sent in fire-ships, which forced the Armada ships to cut their cables and scatter, thus losing their usual effective crescent-shaped fighting formation. Another very damaging attack was sustained off Gravelines on August 8, and Medina Sidonia, realizing that the fleet was in disarray and unable to beat back against the winds, immediately made plans to get the remaining ships home via the north of Scotland and along Ireland's Atlantic coast. He specifically exhorted his commanders "to take great care lest you fall upon the Island of Ireland, for fear of the harm that may befall you on that coast." However, the autumn weather of 1588 was one of the worst on record and the ships, already damaged in battle, were driven off course by the violence of the storms. Over 20 were wrecked along the north and west coasts of Ireland.

Four centuries later, the noted Belgian under-water archaeologist Robert Stenuit, by a combination of academic research and sheer physical effort, was able to reveal the wreck site of the galleass *Girona*, which had not managed to survive the difficult journey home. The *Girona* collection was acquired by the Ulster Museum in 1972 – for the first time, an entire assemblage from an underwater

excavation was purchased by a single institution and preserved for the nation.

The Galleass *Girona*

The *Girona* was captained by Fabricio Spinola of Genoa, and it was one of four galleasses of the Naples squadron. However, at the time of its sinking off Lacada Point, near the Giant's Causeway in County Antrim, on the morning of October 26, 1588, it was actually commanded by Don Alonso Martinez de Leiva. He was a member of one of Spain's noblest families and one of the youngest and most admired of the military leaders. A favourite of Philip II, he was commander designate of the Armada in the event of the death of Medina Sidonia. Having survived the wreck of two other ships, the *Sancta Maria Encoronada* and the *Duquesa Santa Ana*, he heard that the *Girona* was sheltering in Killybegs harbour, Donegal, attempting to make some repairs. He struggled north with the remnants of the crews and infantry of the two ships. Jettisoning most of the heavy cannon on board the *Girona*, a ship designed to hold 550 men became crowded with 1,300. When the *Girona* was dashed to pieces at Lacada Point, only five men survived and the brave de Leiva was not among them.

The first object that Robert Stenuit and his team found was a heavy boat-shaped lead ingot carried by Armada ships to make shot for muskets and arquebuses. Gradually, objects of a much more opulent and glamorous nature began to be recovered.

Renaissance Gold

The sumptuous gold jewellery comprised orders of chivalry, religious orders, and purely ornamental personal decoration, such as gold chains and gold buttons. Hundreds of gold and silver coins were recovered, as well as silver candlesticks and the remnants of silver-gilt tableware designed to grace the table of the aristocratic ship's officers. Philip

Against a background of 1,800 grams of gold chains,
a Jesuit ring and two elements of gold necklaces
Chased jewellery, adorned with a ruby and two pearls (right)
and probably a cameo, now lost (left).

Photograph: Marc Jasinski

II was an austere and devout man, plain in his dress and ornament, but this was not reflected by the members of his court in either garments or jewellery. The Spanish officers on board dressed in the same luxurious splendour as they did when attending Court, complete with gold buttons, gold chains, gold rings, orders of chivalry and glittering diamond rings. The reason there was such wealth on board the *Girona* was because it was carrying the crews, and officers, of two other ships and the leaders of the Armada, who had been chosen from the noblest and wealthiest houses in Spain.

Knights

Many of the jewels reveal facets of sixteenth-century Spanish history. The Orders of Chivalry of Compostela and Alcantara were originally founded in the twelfth and thirteenth centuries as religious military orders, with the purpose of protecting pilgrims to the Holy Lands and recapturing Christian shrines from the Muslims – or Moors, as they

were called at that time. Each knight wore a cross of the saint connected with his order and pledged to drive the Moors out of Spanish territory. In succeeding centuries, orders were awarded to the nobility more as badges of office and to underline their importance in society, rather than for courage and military honour. A gold cross of the Order of a Knight of St John of Jerusalem may have belonged to the captain of the *Girona*. We do not know the owner of the oval doubled-sided Cross of a Knight of Alcantara, which had St. Julian of the pear tree on one face and an openwork gold cross on the other. The exquisitely simple gold cross, inlaid with red enamel, of a Knight of the Order of Santiago

de Compostela belonged to de Leiva. In the Middle Ages, the shrine of St. James of Compostela was an important place of pilgrimage for Irish pilgrims and others.

Religion

Philip II considered the Armada a religious crusade, and forty clerics travelled on board ship. All of the crews were issued a religious medal in copper or pewter. A gold ring with the initials "IHS" on the bezel was found. These initials are an abbreviation of the Greek for "Jesus" and are the sacred monogram of the Jesuit Order. An even more splendid religious jewel is an Agnus Dei reliquary in the form of a small golden book with St. John the Baptist depicted on the front cover. It contained little wax pellets made from Paschal candles mixed with consecrated oil. These had been blessed by the pope at the start of the grand enterprise and were believed to have miraculous powers of protection.

The Jewellery

Twelve gold rings were recovered from the *Girona* and, although most had lost their jewelled settings, one still retained two cut diamonds. The personal and sentimental nature of the gold rings is emphasized by two in particular, the "Madame de Champagney" ring and the "no tengo" ring. Madame de Champagney's ring was being worn as a family heirloom by her grandson, Don Tomas Perrenoto, who was just 21 years old when he perished with the *Girona*. No one knows who owned the little gold ring of a hand holding a heart, with the legend "No tengo mas que dar te" – I have nothing more to give you.

In a much more ornate style were the two surviving segments of a composite gold chain with a central ruby setting, flanked on either side by a pearl. Such grand jewelled chains are often depicted in contemporary portraits of Spanish grandees.

Twelve portrait cameos of Byzantine Caesars, in lapis lazuli, gold, enamel, and pearls, would have formed a spectacularly ornate chain for one of the *Girona's* officers. Two of the most complete cameos still retain some of their green enamel and four pearls on each side. The pearls came from the oyster beds of Venezuela, traded by Indian pearl divers with the Spanish conquerors. Only 11 of the cameos were found during the original excavations, and Robert Stenuit maintained that there should have been a twelfth to complete the set. In 1997, almost 30 years later, a diver made the remarkable discovery of the twelfth!

Two other jewels – a gold salamander set with rubies and a gold ring with a salamander on the bezel – reflect the grandeur of the Spanish empire. Gold and silver from the Spanish conquests in the Americas were a primary source of income for the empire. While large numbers of splendid native gold ornaments were melted down for bullion, the ideas of South American craftsmen were often retained. This may explain the fondness for jewels in the form of animals, which appeared during the sixteenth century. Hernan Cortes, the conqueror of the Aztec Empire of Mexico, recorded in 1526 that among the gold ornaments sent to Spain was a "winged lizard" or salamander. In legend, the salamander was believed to have the magical properties of being able to extinguish and survive fire – this made it a potent good-luck charm on board a wooden fighting ship, on which fire was one of the greatest hazards.

◄ **Gold salamander-shaped pendant inlaid with rubies**
According to legend, the salamander had the magical power to live in and put out fires, which made it a powerful talisman on board wooden warships, where fire was the worst danger.

Sixteenth century
Ulster Museum, Belfast, inv. BGR1
Photograph: Michael McKeown/The Ulster Museum

▲ **Gold Spanish escudo bearing the coat of arms of Philip II on its obverse**
The top of the blazon represents the coast of arms of his parents, Charles Quint and Isabelle of Portugal, while at the base are the arms of Portugal affirming his reign.

Late sixteenth century
Photograph: Michael McKeown/The Ulster Museum

Archaeologist and diver admiring their discoveries:
a handful of gold ducats and an 800-gram gold chain.
Photograph: Marc Jasinski

The salamander appears again on the bezel of a fine gold ring, where it is flanked on either side by a human head. The head was frequently depicted by pre-Hispanic Indian goldsmiths on breast, nose, and ear ornaments.

Six gold chains were recovered from the galleass, two of which were of large and heavy plain links. It is recorded that each Armada captain wore a heavy gold chain over this undershirt and under his doublet. The four other gold chains were fragmentary, much finer and more intricately made, but all reflected the wealth and grandeur of their owners.

The Coins

The hundreds of gold and silver coins represent the personal wealth of those on board, but they also emphasize the close connections among medieval Spanish wealth, Spain's conquests in the New World, and the extent of its empire in the sixteenth century. The gold, silver, and copper coins had been minted in six different countries: Spain, Portugal, the Kingdom of the Two Sicilies, the Republic of Genoa, Mexico, and Peru. Although by 1588 the Spanish had established mints in Mexico, in Lima, Peru, and in Potosi, Bolivia, almost 85% of the gold coins found had been minted in Seville. This important and wealthy city of southern Spain was the centre for gold coming from the New World.

A single Portuguese gold coin, a Sao Vicente from the Lisbon mint of the Portuguese king, John III (1521–57), is a further reminder of the power of Spain's empire in the sixteenth century. Its conquest of the Portuguese empire gave it access to the wealth of the Portuguese colonies and the very profitable trade routes.

In more recent years, the Ulster Museum acquired a pendant jewel that provides an epitaph for the thousands of Spanish soldiers and sailors who perished on the shores of Ireland in the aftermath of the Armada. The centre of the pendant is a gold and blue enamel earring of the Virgin and Child. It was customized later with the addition of a gold border set with an octagonal amethyst and four emeralds and hung with a pendant pearl. The verse round the border, in medieval English, reads,

> When Spanneshe fleet fled home for feare
> This golden picktur then was found
> Fast fexed unto Spanniards eare
> Whoo drowned lay on Irish ground Anno 1588.

Portuguese gold coin struck in Lisbon
in effigy of St. Vincent (São Vicente)

Late sixteenth century
Ulster Museum, Belfast, Inv. BGR 800
Photograph: Michael McKeown/The Ulster Museum

The Treasures of the Spanish Galleons on the Maritime Routes

LETIZIA ARBETETA MIRA

Museo de América de Madrid

Translated by Käthe Roth, from the French translation by Louis Jolicœur

THE PARTICULARLY EVOCATIVE SUBJECT of gold has already been written about extensively. Contemporary historiography arising from the nineteenth century, has, however, greatly reduced the role of the Iberian nations – Spain and Portugal – in the complex history of the precious metals of the Americas. By combining contemporary political concepts and attitudes with historical reality, the gold–conquistador–pillaging equation has become one of the most deeply rooted common ideas in modern America. This common idea also touches on the very nature of the Spanish presence on the continent, and in some way critiques it.

In the second half of the fifteenth century, two of the three great states on the Iberian peninsula, Portugal and Aragon, had found a solution to their need for gold, which was an indispensable part of each state's economic structure and credibility. Both nations had guarantees of trade expansion – for Aragon, by access to the Mediterranean Sea; for Portugal, via the sea routes off Africa leading to Asia. These routes had been established thanks to successive "discoveries" made by Portugal, and their exclusive use had been granted by the papacy in its role as international arbiter. Squeezed between these giants, a third state was trying to make a place for itself: Castile. The routes had been defined by treaties such as the one signed at Alcaçovas in 1479, which forbade navigation by other nations in waters conceded to Portugal. Although Castile had different points of access to the sea, it had an urgent need for privileged maritime trade routes, as well as for means of obtaining gold on a regular basis. This is why the Crown could not turn down an offer such as that by Columbus, who proposed to find a direct route to India, the land of spices and marketplace for Asian products.

When Admiral Columbus, commanding an official expedition composed of excellent sailors and workers, touched land, he believed that he was setting foot in India. Castile immediately claimed this new route, on the pretext that it had discovered new ports and wished to fortify them to ensure the protection of its trade. The expedition had another important economic mission: to check for the presence of gold, worked or not. The exploitation of gold deposits thus came under the prerogative of the Crown, and worked gold was to become an object of conquest or trade. Soon, Spain (created by the merger of the crowns of Castile, Aragon, Navarre, and other territories) was speaking of a new continent, while Portugal claimed a part of the discovered land, on the basis of previously conceded rights. The Treaty of Tordesillas (1494) traced a line marking out the territorial possessions of the two countries.

In the meantime, trade structures were becoming organized in Seville, one of the largest cities on the Iberian peninsula and a flourishing trade centre. Unlike the Portuguese model, in which trade was monopolized by the Crown, Spanish trade remained free for Castilians. Only certain goods were reserved for royalty, including the precious metals extracted from its mines.

Oceanica Classis

America, called "the West Indies" by the Spanish, was not considered a colony. It was part of the Hispanic kingdom, and was therefore subjected to general laws and certain special laws (Law of the Indies), under the authority of two viceroys ruling, respectively, New Spain to the north and Peru to the south. (The term "colony" applied to America by authors of treaties appeared only in the eighteenth century.) Placed under Spanish administrative control, representatives of the viceroyalties avoided severe punishment for abuses when, in the seventeenth century, administrative corruption began to be propagated from the homeland.

Large fleets were assembled to fend off attacks by foreign ships. These were, in fact, expeditions protected by armed ships, grouping together official and trading vessels. In the sixteenth century, ships' holds were filled mainly with the European goods needed to build a new society. In America, these goods were paid for either in cash or in the form of barter for other novelties: plants for dyes and medicines, exotic woods, various raw materials, and manufactured products.

Thus, a distinction must be made between trade and official extraction of precious metals. Gold was rapidly overtaken by silver mined from large deposits such as those on the hillside in Potosí, Peru, and the town of Guanajuato in Mexico. So much gold was mined that the global price dropped and silver became the distinctive sign of the noble class of the *Indianos* (emigrants who returned to Europe after enriching themselves in America). Although there are diverging opinions with regard to the final destination of these huge quantities of silver, the dominant theory is that it was used to strike coins for the Asian markets; countries such as China required payment for their products in silver coins, while gold was used as a bank guarantee in Europe. Some of the silver was also used to decorate furniture in private houses and churches

and to make donations, considered a symbol of prestige in Spanish society. Despite the widespread pillaging and destruction resulting from the wars of succession from 1702 to 1713, the Napoleonic invasion, and the civil war of 1936–39, Spanish churches still contain numerous objects in worked silver from the America of the Spanish viceroys.

Over time, ship owners and merchants tended to choose to transport luxury products of small size and great value, due to the very high cost of maritime voyages and their duration (between one and six years), and in response to demand from European and American markets. Fraudulent declarations about the value of products were, however, so common that it is impossible to know the real volume of commercial traffic, which carried not only luxury items manufactured in Spain, but also foreign products introduced into Spain in different ways. The prices of high-end fabrics, ribbon, embroidery, notions, garments, jewellery, shoes, hats, and other luxury products climbed to unprecedented heights in both viceroyalties. Because many of these articles came from France, a number of Spanish traders posed as Frenchmen to sell them at a better price. Linens, mattresses, bedclothes, and domestic fabrics were also greatly prized, although the opening of the Asian market, thanks to galleons sailing between the Philippines and Mexico (from Manila to Acapulco), now provided access to fine silk and cotton fabrics from India and China. Books, a good number of which were printed in the Netherlands, were heavily censored. This was also the case for engravings, which were used as models by artists, and sculptures and paintings by Spanish masters and some foreigners, such as Zurbarán, who had specialized studios producing works for export. Food products, wines and spirits, iron tools, machinery, and other instruments also filled ships' holds. Some wealthy people had large cargos of furnishings brought to make their lives more pleasant.

America imported raw materials newly in demand, such as cacao, Asian products such as porcelain, silk, ivory carved to suit European tastes, and gold and silver worked in various techniques and motifs. The Spanish galleons transported to Europe marvels such as the "Pilgrim," the largest pearl ever found, and a series of large emeralds from Chivor and Muzo in what is today Colombia,

◄ Oceanica Classic, **Woodcut engraving of a Spanish galleon**
Portrayal of Christopher Columbus's (1451–1506) caravel the Santa Maria. Engraving from "De Insulis in mare indico super inventis," Basle 1493–94 (Latin translation of Columbus's report to the king of Spain).

Photograph: © Costa/Leemage

Jus. Castellanos f.te. mex.°   año. de 175.

including the one in the Imperial Chamber in Vienna that weighs 2,680 carats. As for gold jewellery, the styles so carefully copied Spanish models that it was difficult to tell American production from Spanish. The pieces recovered from shipwrecks do not help to clarify the question, since they may have been personal objects from different eras and places.

Silver and gold belonging to the Crown, import duties such as the famous Quinto Real (20 percent of the value of the merchandise), fines, and various donations were used to build routes, fortifications, and buildings of all types in American towns, as well as paying the salaries of functionaries and soldiers; the costs were sometimes higher than the profits.

To return to private trade, in the sixteenth century traders began to ship some curiosities, such as Mexican feather mosaics portraying religious themes, natural science specimens destined for curiosity cabinets, and a good number of collector's objects, some dating from the pre-Hispanic era. Among the luxury items, other than articles from Asia,

furniture was particularly prized, as were objects made in America from raw materials, such as chests decorated with shells, and sculptures made of ivory, alabaster, or wrought silver using particular pre-Hispanic techniques.

Because the Spanish particularly appreciated religious objects such as public or private identitary symbols, artisans began to paint effigies of famous virgins such as the Virgin of Guadalupe, images of Christ, and saints venerated in America and to sculpt votive statuettes. There were also some eccentric paintings, such as the series dubbed "from the land" and painted expressly for export, evoking the mixing of blood with a touch of irony and exoticism, also very much in fashion for curiosity cabinets. Other manufactured products transported in ships also responded to the infatuation with exotic articles: mainly boxes, chests, and trunks of different shapes and materials, folk-art figurines, and objects with mother-of-pearl inlay.

Some cities located on the trade routes linking East and West, such as Puebla de los Angeles in Mexico, became wealthier than the homeland. However, they declined little by little, undermined by wars and independence movements. The most important fact about this adventure is that from the sixteenth to the nineteenth century, Spanish ships transported in their holds for trade the treasures of three continents.

Chalice made of gold and silver inlaid with diamonds, rubies, and emeralds, from South America, 1768

Museo nacional de Arte Antiga, Inv. 346 OUR, H: 28.5 cm
Photograph: © José Pessoa, Museo nacional de Arte Antiga/Divisão de Documentação Fotográfica – Instituto dos museus e da concervação, I.P.

◄ Virgin of Guadalupe

Highly venerated in South America, the Virgin of Guadalupe appeared to an Indian in 1531 and miraculously answered his prayer, which earned her an almost immediate cult following. It is said that the Virgin of Guadalupe protected the Indians against massacres, disease, and natural disasters.

Fabric embroidered with gold and silver, paper, and polychrome wood
Juan Castellanos, seventeenth century, H: 61.5 cm W: 46.5 cm
Museo de América, Madrid, Inv. 1987/01/01. Photograph: Museo de América

CHAPTER **4**

ARTISANS OF GOLD

The Origin of Metallurgy in North America

CLAUDE CHAPDELAINE

Archaeologist, Professor, Department of Anthropology, Université de Montréal

Translated by Käthe Roth

THOSE WITH AN INTEREST in the origins of metallurgy in the Americas usually look to South America or Mexico, where the most impressive pre-Columbian civilizations developed, to find the earliest evidence. Although the idea of a number of independent, distinct points of origin in the Americas as a whole is a valid one, the oldest point of origin is situated in North America. A North American metallurgy, independent of the great civilizations, developed around the Great Lakes more than 6,000 years ago. Copper, which was abundant on the surface (unlike gold), was at the centre of this innovation.

The first North American metallurgists were hunter-fisher-gatherers who moved around their territory following the seasons. Their technology was dominated by utilitarian chipped and drilled, polished tools. They made stone spearheads, knives, scrapers, and drills, and polished various materials to shape axes, adzes, gouges, a few pendants, and even points. To this panoply of stone tools were added others made of bone, such as fishhooks, harpoons, awls, points, and needles. The copper objects at first imitated various common objects that had been chipped, polished, or shaped out of bone or antler. The addition of copper tools was thus complementary to these nomads' toolkit. It is even possible that copper was tested for its

Locations of copper deposits around Lake Superior and archaeological sites on Allumettes and Morrison islands in the Outaouais region

Photograph: Claude Chapdelaine

effectiveness and durability compared to other tools with a similar function made from materials that tended to be less expensive, more accessible, and lighter. Copper fulfilled a utilitarian function for more than 3,000 years, and populations with direct access to copper-bearing regions were to develop an expertise that was manifested in the great diversity of tools produced, the main types being points, punches and awls, fishhooks, and a few large-sized tools such as axes and gouges. The great majority of copper tools were small and made from copper sheets and rolled copper, except for large points, axes, and gouges. We can thus speak of a thrifty copper industry.

It is possible to imagine that following the discovery on the surface of nuggets of copper of different sizes in a number of regions around Lake Superior, curious individuals, attracted to the metal's reddish colour, gathered some pieces and processed them by cold hammering. The metal may first have been used as a flint, due to its weight; with observation of how its surface heated in repeated contact with a flint stone, the idea of processing it may have come about. The first step seems to have been repeated hammering to flatten the mass. The goal was to obtain a sheet, which could then be folded and rolled to make a spindle. A number of tools were made from these spindles. To this day, cold hammering is the dominant technique for processing copper, but there are many indications that it was often heated to make the work easier. However, it was never

◄ Glaciers, 1990

Photograph: Pierre L'Heureux

Large copper arrowheads found in Outaouais

Photograph: Claude Chapdelaine

heated to a liquid state by peoples in the Great Lakes region. The technical prowess of the first metallurgists went only as far as cold working and tempering.

Very soon after copper was discovered and began to be transformed into common objects, the neighbours of the first metallurgists became covetous. Copper objects quickly became valuable trade goods. There are a number of reasons for this infatuation. First, the red colour of the metal was attractive; second, copper was rare. Its limited geographic distribution was noted by Jacques Cartier during his voyages; he located its source in the mythic Saguenay kingdom. In fact, relying on what his Aboriginal interlocutors told him, Cartier was indicating that the source of the metal was located to the west, in the Lake Superior region, not around what is today called the Saguenay–Lac-Saint-Jean region.

It is not known whether demand for copper objects grew due to their efficiency, but, except for two sites located in the middle of the Outaouais Valley region, red-metal tools remain very rare in assemblages. However, they continued to circulate until the historical era. The Outaouais region was probably the copper route linking regions farther west to the St. Lawrence Valley. Two exceptional sites were excavated in the early 1960s; the analyses of these collections were recently published, lifting the veil on how copper circulated (Clermont and Chapdelaine 1998; Chapdelaine and Clermont, 2006; Clermont, Chapdelaine, and Cinq-Mars, 2003). These two sites are on Morrison and Allumettes islands; hundreds of copper objects, of a very wide variety, were found there. New radiocarbon dating on bone remains on these sites confirms that the Allumettes Island site is the older one, and that groups in the Outaouais region

quickly became involved in trade with groups in the Great Lakes. In fact, the dates obtained are as old as the dates recorded in the Great Lakes, indicating that copper goods were quickly integrated into a vast network of interactions, almost immediately after metal-working began.

In addition to revealing the undeniable cultural links with metallurgists in the environs of Lake Superior, the great quantity of metallic waste in the sites on Allumettes and Morrison islands no doubt indicates the presence of workshops. Although the majority of copper tools circulated in a finished form, the occupants of the sites in the Outaouais knew how to work the metal. There was a desire to use and maximize this rare material. Thus, metal spindles were regularly modified to produce regular fishhooks, composite fishhooks, punches, awls, and needles. The popularity of copper is undeniable; the collections from the Outaouais support the idea that small, easily transportable objects were circulated, but also a thrifty use of the material, and massive objects remain extremely rare.

After a "golden age" of copper that lasted several millennia, about 3,000 years ago utilitarian tools made of copper became less abundant. The production of copper objects continued, although in a limited fashion, and non-utilitarian objects, such as bracelets and necklaces, became more sought after. Copper objects were no longer in competition with stone objects with similar functions. Rather, copper took on a strong symbolic function, as a funerary offering or simply as an appreciated gift to seal an alliance.

Bibliography

Chapdelaine, C., and Clermont, N. 2006. "Adaptation, Continuity and Change in the Middle Ottawa Valley: A View from the Morrison and Allumettes Island Late Archaic Sites." In David Sanger and Priscilla Renouf (eds.), *Archaic of the Far Northeast*, pp. 191–219. Orono: University of Maine Press.

Clermont, N., and Chapdelaine, C. 1998. *Le site archaïque de l'île Morrison.* Paléo-Québec coll., No. 28, Recherches amérindiennes au Québec, Montreal.

Clermont, N., Chapdelaine, C., and Cinq-Mars, J. 2003. *L'île aux Allumettes et l'Archaïque supérieur dans l'Outaouais.* Joint publication. "Paléo-Québec" Collection, No. 30. Montreal and Hull: Recherches amérindiennes au Québec and Musée canadien des civilisations.

The Metallurgy Practised at the Huacas de Moche Site

CLAUDE CHAPDELAINE

Archaeologist, Professor, Department of Anthropology, Université de Montréal

Translated by Käthe Roth

Crown

This crown, a masterpiece of Chavín goldsmithery, is adorned with a relief portrayal of the god of sticks wearing a snake belt, with his traditional feline canines and clawed hands and feet. He is flanked on either side by characters in profile with bird's talons, in whose mouths are winged animals.

Chavín, Sierra Norte, Valle del Mosna, Chavín de Huantar, Peru
900–200 BCE
Hammered gold
H: 18.1 cm W: 44 cm D: 0.1 cm
Museo Arqueológico Rafael Larco Herrera, Inv. ML100541
Photograph: Museo Arqueológico Rafael Larco Herrera, Lima, Peru

THERE ARE A NUMBER of criteria for defining a complex society. Aside from monumental architecture, the recognition of a class of specialized artisans, more particularly those who work with metals, is a major factor. The practice of metallurgy developed slowly in South America, with its origins going back to the early second millennium BCE. More sustained experimentation developed during the first millennium BCE, resulting in the creation of the first gold-leaf hammered masterpieces of the Chavín culture. During the reign of the Moches, between 100 and 800 CE, specialized artisans worked full time producing objects in gold, silver, and copper. These metallurgists, working mainly for the elites, invented a new method for responding to the growing demand; they covered copper objects with a thin film of gold, thus giving the impression that the objects were made of solid gold. It was probably the weight of these objects, called *tumbagas,* that betrayed their true nature. This method was intended to satisfy growing demand and limited supply due to a relatively small number of easily accessible gold deposits visible on the surface of the ground. The combination of copper and gold or, more rarely, silver became the trademark of the metallurgic production of the Andean area up to the arrival of the Spanish. The production of numerous *tumbagas* did not interfere with the production of solid-gold objects, but the latter were rarer and no doubt had grater symbolic value. Although there was no "golden age" of production

of gold pieces during the Moche civilization, the leaders of this nation on the north coast of Peru played a decisive role in the blossoming of great artistic diversity. How were these pieces produced? The answer to this question takes us to Huacas de Moche, the name given to the site of the capital of the first South Moche state. There, archaeologists have had a chance to excavate a unique structure in the Moche world, a kiln with chimney used to process metal.

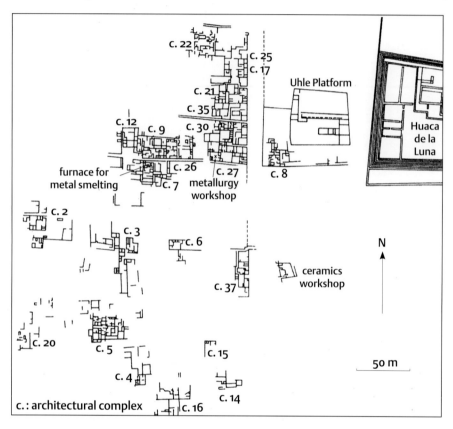

Plan of the urban zone west of Huaca de la Luna, Huacas de Moche site, and location of the two workshops associated with metalwork

Photograph: Claude Chapdelaine

A Capital, a City, and a Production Site

The Moche civilization developed on the north coast of Peru at the beginning of the first millennium BCE, and its apogee extended over a number of centuries, from 350 to 750 CE. Archaeologists now divide the Moche world into two major cultural and geographical spheres that had a certain number of features in common and others unique to each. In the southern sphere, political power was more centralized and large urban centres developed in the Chicama and Moche valleys. This

is where the largest site, Huacas de Moche, with an area of at least sixty hectares (one kilometre long by six hundred metres wide), is located. Two temples, or *huacas,* have been found, constituting the main monumental manifestations of the new central power; between the two public buildings was a zone containing many residences of the elite class (Uceda, 2001). The large size of the two temples, built of bricks made of sun-dried mud, is the main factor supporting the hypothesis that the Huacas de Moche site was the capital of a state organization. The layout of the city is also very complex, with temples, *plazzas,* and residences organized around a network of streets and alleys; this urban plan is surprisingly modern and has not been seen in other valleys dominated by the Moche.

In this capital of an expansionist state, the urban zone was not solely residential in nature. Recent excavations have established beyond doubt the importance of craft production. A wide variety of other economic also took place there. Excavations have revealed the existence of two workshops producing ceramics, small lapidary workshops, food-production areas (mainly maize beer), cotton-spinning areas, weaving centres, and two metallurgy workshops.

The Metallurgy Workshops

During archaeological excavations of the last ten years, the discovery of a number of workshops has revealed an unsuspected complexity and confirmed the importance of craft production at the heart of the capital. The specialized artisans lived under the patronage of the leaders. To date, two areas excavated in the urban zone have been interpreted as being workshops for metalworking. The first, in the north part of complex No. 7 (Chapdelaine, 1997: 50–52; Chapdelaine, 2002: 70–71; Chapdelaine et al., 2001: 388–89), has a kiln, while the second is associated with complex no. 27, closer to the Temple of the Moon. These two workshops probably fulfilled different functions, with the first being used for smelting, while the second had a panoply of tools used for the production and finishing of metal objects. The workshops are rather small in size, but the pace of work must have been steady. There were other workshops located around the urban area, indicated by the discovery of a number of ceramic-blowing tubes distributed on the surface. The blowing tubes were at the end of long reeds into which the artisans blew to maintain constant heat in the kilns. The location of these workshops in the heart of the urban centre is surprising, given that fire and its consequences could be disastrous for the neighbours. Their central position was certainly justified by the control that the leading class wanted to exert over production of highly desired and prized metal objects. It must be remembered that metals were rare in the region immediately around the site and were imported from the Andes. The limited availability of metals, copper as well as gold, explains in large part the fact that production of non-utilitarian objects was reserved for the elites.

Vestige of a Moche furnace

The chimney was 1.20 m high; it was composed of nine rows of adobe; the internal diameter is 55 cm; and the exterior base measures 15 x 15 cm.

Photograph: Claude Chapdelaine

A Unique Structure

In the summer of 1996, there was a surprising discovery: a kiln with chimney within a residence in the centre of the urban zone between the two temples. There was thus metal being smelted in a workshop attached to a residence in the elite area, judging by the quality of the construction, the size of the rooms, the number of storage rooms, and a maize-beer-production zone. The chimney, situated in complex No. 7 and carbon-14 dated to between 435 and 665 CE, was used during a period corresponding to the apogee of the Moche capital.

The circular chimney was built with eight superimposed rows of dried-mud or adobe bricks. It had an inner diameter of 55 centimetres and a height of 1.2 metres over a clay floor. When the outer wall was uncovered, it was found to have a small quadrangular orifice measuring 15 cm², near the base, that was blocked by a crudely sculpted adobe brick. On the inside, at the very bottom, some slag and charcoal covered the hardened floor. The inner perimeter of the chimney was coated with clay heavily rubefied by heat. Based on the presence of a ceramic crucible covered with copper oxide, found nearby, and a significant concentration of copper objects in the sector, we concluded that the kiln at the base of the chimney may have

been used for smelting metal. We therefore sampled the clay coating in the chimney and used neutron activation analysis to determine its chemical composition. The result indicates an unequivocal contamination of the clay with a concentration of gold sixteen times greater than in the other clay studied on the Huacas de Moche site (Chapdelaine et al., 2001: 388). There is no doubt that the facility in this workshop was used to smelt metal and that this metal contained gold.

Neutron activation analysis of a number of copper objects in the urban zone also demonstrated a significant presence of gold among the chemical elements identified. Deterioration of the thin gold film kept the archaeologists from recognizing other objects that had been covered with gold within the site. These results confirm the importance of *tumbagas* in Moche metallurgy. In addition, on a more technological note, the results of the neutron activation analysis also confirmed that making a deliberate alloy of copper and arsenic to produce an early bronze was rare (Chapdelaine et al., 2001: 389). Several centuries later, this development was the prerogative of metallurgists of the Sicán culture, situated farther north in the Lambayeque region (Lechtman, 1996; Shimada and Merkel, 1991).

Conclusion

The discovery of this chimney helped to provide a new perspective on the Huacas de Moche site, now considered a place of production of precious objects made of gold and gold-leafed copper. It also renewed research on the Moche metallurgic tradition. The chimney is unique, and its shape contrasts with a depiction of metallurgists at work on a Moche ceramic vase (Donnan, 1978: 12), showing three people blowing into long tubes introduced into a ditch-shaped kiln. Our chimney is, rather, similar to a form noted in the early seventeenth century and identified as being for smelting metal (Ravines, 1978: 479). Specialized analysis of clay samples taken from inside the chimney supported the hypothesis that gold had been smelted in this kiln and that other cleaned metals, such as copper, were also smelted there. The result is a new way of looking at the metallurgy in the Huacas de Moche site, as well as confirmation of the importance of

Vase illustrating a scene of metallurgy work
Photograph: Christopher B. Donnan

gilded copper objects, the *tumbagas* that became the norm rather than the exception for the great urban class. The few objects in solid gold were reserved for the supreme leaders.

Bibliography

Chapdelaine, C. 1997. "Le tissu urbain du site Moche, une cité péruvienne précolombienne." In C. Chapdelaine (ed.), *À l'ombre du Cerro Blanco. Nouvelles découvertes sur le site Moche, Côte nord du Pérou*, pp. 11–81. Les Cahiers d'anthropologie, No. 1, Université de Montréal.

———— 2002. "Out in the Streets of Moche: Urbanism and Socio-political Organization at a Moche IV Urban Center." In W. Isbell and H. Silverman (eds.), *Advances in Andean Archaeology and Ethnohistory*, pp. 53–88. New York: Plenum Press.

Chapdelaine, C., Kennedy, G., and Uceda, S. 2001. "Neutron Activation Analysis of Metal Artifacts from the Moche Site, North Coast of Peru." *Archaeometry*, 43 (3): 373–91.

Donnan, C. B. 1978. *Moche Art of Peru*. Los Angeles: Fowler Museum of Cultural History, UCLA.

Lechtman, H. 1996. "Arsenic Bronze: Dirty Copper or Chosen Alloy? A View from the Americas. *Journal of Field Archaeology*, 18: 43–76.

Ravines, R. 1978. *Tecnologia andina*. Lima: Instituto de Estudios Peruanos.

Shimada, I., and Merkel, J. F. 1991. "Copper-Alloy Metallurgy in Ancient Peru." *Scientific American*, 265: 80–86.

Uceda, S. 2001. "Investigations at Huaca de la Luna, Moche Valley: An Example of Moche Religious Architecture." In Joanne Pillsbury (ed.), *Moche: Art and Political Representation in Ancient Peru*, pp. 47–68. Washington, DC: National Gallery of Art.

The Gold of the Andes:
From Decoration to Divine Power

CLAUDE CHAPDELAINE

Archaeologist, Professor, Department of Anthropology, Université de Montréal

Translated by Käthe Roth

THE TREASURES OF PREHISTORY are always fascinating, and the discovery of gold objects, whether they come from the steppes of Siberia, the banks of the Nile, or the north coast of Peru, is a direct reminder of the value that we accord to this precious metal, the universal synonym for wealth.

Forehead ornament decorated with a human head wearing a coverchief in the form of a half-moon and two zoomorphic animals representing dragons with feline heads and dentate bodies.

Mochica, Costa Norte, Peru, 200 BCE– 600 CE
Museo Arqueológico Rafael Larco Herrera, Inv. ML100769
H: 26.6 cm W: 29.5 cm
Photograph: Museo Arqueológico Rafael Larco Herrera, Lima, Peru

Artisans of Gold

Pair of earrings
Quimbaya, Colombia, 400 BCE–1600 CE
Museo del Oro, Inv. O 06475, O 06476, Diam: 7.7 cm
Photograph: Clark Manuel Rodriguez,
Museo del Oro, Bogotá, Colombia

However, not all civilizations of the past have had the same idea about gold. More than 3,000 years ago, South American peoples began to hammer gold to transform it into leaf, and their interest in metals never subsided. Metals became important as a factor of social differentiation; notwithstanding stories of untold wealth in gold and silver in Peru, these metals were not very abundant, and their rarity made them precious. Many sources were situated at a high altitude, in places distant from the large urban centres. Although gold and silver were already considered precious metals, just as they are today, in that ancient era copper was also considered a semi-precious metal. Its relative rarity and limited distribution on the surface or in rocky outcroppings support this categorization. The continuous production of non-utilitarian objects reserved for elites is probably explained by this rarity, especially with regard to copper, which was rarely used to create tools for daily tasks. Metal objects and, in particular, those made of gold, always were luxury items.

A Brief History of Metallurgy

Metallurgy has a long history in South America. It began around 500 BCE in the Peruvian Andes. There is little evidence, but the cold hammering of gold, then of copper, seems to have been the first step during the Initial Period of 2000 to 1000 BCE. In the Early Horizon period (1000 to 200 BCE), objects made of shaped gold became more common in the Andes and on the Pacific coast. They were decorated and bore stylized messages of Chavín culture (Burger, 1992; Onuki, 2001). One might infer that they were made by specialized artisans.

Not until the Early Intermediate period (200 BCE to 600 CE) was there spectacular progress in metallurgy on the north coast of Peru. The Vicú and Moche cultures developed new production and decoration techniques to satisfy the growing demand for metal objects. It was the beginning of

a new age of metals, that of alloys.

The rarity of gold and the increasing demand forced metallurgists to use copper, which was more abundant, as a base and cover it with a thin film of gold or, occasionally, silver. These gold-leafed copper objects were named *tumbagas* and they were made to fulfil the extravagant needs of the elites, who increasingly coveted valuable goods to distinguish themselves socially and also to accompany them on their voyage to the hereafter. As the Moche leaders, who probably ruled the first nation on Peruvian territory, consolidated their centralized power, there seems to have been strong growth in the use of this medium to convey an ideological message. A variety of decorative techniques were employed to raise metal objects to the level of the terra cotta masterpieces made by extraordinarily creative Moche ceramicists.

Productivity in the metal arts reached its apogee between 900 and 1300 CE in the Lambayeque region, on the north coast of Peru. The Sicán culture was behind this development. Metallurgists in the region regularly used a new alloy to improve and strengthen copper: they added enough arsenic to create bronze. Later, the Incas added tin to copper to create a second type of bronze. The Sicán specialists produced an impressive quantity of gold-leafed copper objects. Pure-gold objects were increasingly rare, and *tumbagas* were omnipresent in this metallurgic tradition. A particular style developed in the region around Batán Grande, the first Sicán capital. These objects have been found throughout much of what is now Peru. The Sicáns fell to the Chimús around 1350 CE, and the conquerors quickly integrated the Sicán specialists into their own workshops. A similar fate was in store for the Chimú metallurgists and the descendants of the Sicáns when the Incas conquered the territory around 1470 CE. Many historical texts note that the Cuzco goldsmiths at the time of the Spanish conquest were originally from the north coast of Peru.

Although Peruvian metallurgy seems to have developed in the Andes (Bruhns 1994: 174), the north shore of Peru was where it became popular and reached an artistic and technical peak. All of the great Ecuadorian, Bolivian, Peruvian, Colombian, and Chilean civilizations developed a pronounced taste for gold objects. The Nasca (south coast of Peru), Tiwanaku (around Lake Titicaca), and Wari (central Andes) civilizations also valued gold, but it is difficult to gauge the true importance of metallurgy in these civilizations. The archaeological evidence is less bountiful, for their capitals were systematically pillaged, especially the tombs of the elites. At the Huari (Wari) site, the royal tombs discovered in the 1960s and 1970s were completely emptied of their contents.

The Incas inherited a 3,000-year-old metallurgic tradition. Unfortunately, the splendour of this production is no longer accessible today, because the conquistadors melted down hundreds of tons of gold, silver, and *tumbaga* objects to facilitate transportation of metal to Europe and Spain (Ramirez, 1996: 121). The pillaging of archaeological sites since the colonial era, with a new wave during the twentieth century, also stole a great number of metal works from posterity.

In the light of discoveries in Peru over the last 20 years, it is possible to discern three great periods of Andean metallurgy. First, there was production of early objects with a high symbolic content from the Chavín culture. Then came the golden age of artistic production under the Moche, as found in the royal tombs of Sipán (Alva and Donnan, 1993), the discoveries at Huacas de Moche (Uceda, 2001) and El Brujo (Galvez and Briceño, 2001), and the tombs of the lords of Dos Cabezas (Donnan, 2003). The third period is that of the Sicán culture. The discovery of a royal tomb at Huaca Loro, an imposing platform situated in the archaeological complex of Bátan Grande, has confirmed the exceptional quality of the work by Sicán culture metallurgists (Shimada, 1995). The impressive quantity of objects recovered in this tomb provides evidence of the great production capacity of the artisans of this culture.

In Ecuador and Colombia, metallurgic traditions emerged at different times in the first millennium CE (Townsend, 1992). Metallurgy, once established in Colombia, extended northward, where its effects were felt in Central America and the west coast of Mexico. These regions instituted a vast network of interactions that facilitated the circulation of objects, ideas, and certain individuals (Quilter and Hoopes, 2003).

The Aesthetics and Symbolism of Gold

Gold starts as a little pebble, nugget, or a glint in a mineral mass. In their raw state, nuggets and veins have no aesthetic value. On the other hand, gold may have an intrinsic economic value, since it is distributed very unequally and often in remote places. The aesthetics and symbolism of gold are notions that different human groups developed differently. Under the artisan's hammer, gold was shaped into many finished products, transforming the nugget into something else completely. The first masterpieces exhumed in archaeological contexts, crowns

Pectoral with anthropomorphic and zoomorphic figures

Tairona, Colombia, 100–1700 CE
Museo del Oro, Inv. O 14618
Photograph: Clark Manuel Rodriguez/Museo del Oro, Bogotá, Colombia

and breastplates, came from tombs. These early gold objects, associated with Chavín worship, have an undeniable symbolic meaning. As a medium, gold carries an incontestable significance, and its decorative use, a sort of added value, provides a better understanding of its role in the expression of a system of beliefs. It was part of a complex system of representations in the first great Peruvian civilization.

Few iconographic elements were unique to gold objects. There do not seem to have been representations or messages linked solely to gold and its use. On the surface of gold objects were reproduced the same symbolic and artistic messages regularly found in ceramics, textiles, wood, and bone. A number of gold objects, such as breastplates, may have been worn daily, conferring greater importance on those who wore them.

On the north coast, the production of gold objects came fully into its own during the first millennium CE. During this "golden age," the main cultures were Moche and Sicán. Metallurgy developed strongly among different civilizations in Ecuador, particularly the Tolita, and in Colombia among the Quimbaya, Tairona, and Muisca. To understand the many functions of gold in Andean societies, one must first examine the qualities of this metal, which, contrary to current conceptions, had no monetary value.

The Qualities of Gold

To understand the importance of gold, one must understand its rarity or, rather, its reduced availability and its limited and very unequal distribution in the Andean area. Its rarity made it a precious good. It was very malleable, and it could be alloyed with silver, copper, and bronze. The technique of colouring made it possible to gild the surface of copper objects with a small amount of gold; colouring involved using an acid such as urine to remove the copper oxide and then heating the object repeatedly to obtain the desired colour.

An undeniable and primordial virtue of gold was its colour. The bright yellow had a particular significance. Among the Incas, gold was associated with the solar disk; silver, with the Moon. In the Andean imagination, gold was an agent of the Sun, united with the gods, and transmitted its light, heat, and life and fertility.

The lustre of the smooth surface of gold was another quality associated with purity and beauty. On the technical level, the greatest quality of gold was its malleability, although the other qualities exploited judiciously by metallurgists should not be underestimated: its great ductility, useful for making gold wire, and its extensibility, useful for producing large sheets by hammering, a technique that typifies the Andean metallurgic tradition. This extensibility explains the propensity of certain Moche and Sicán artisans to produce masks, crowns, and large hip protectors.

Gold also had some not-so-attractive qualities. Due to its fragility, it never competed with copper and other hard materials, such as wood and bone, in the making of utilitarian tools. Gold objects could not be used for purposes other than those linked specifically to ceremonial functions.

Duality and Political Function

The use of gold was popular and seems to have been widespread. However, in a world where demand was greater than supply, it is understandable that the lower levels of Andean societies did not have access to precious gold, silver, or shell objects from Ecuador.

Gold objects were always associated with luxury, and thus were opposed to utilitarian objects. Because distribution of gold pieces depended on the number of pieces made, gold objects conferred more prestige on those who possessed them.

On the social, economic, political, and ideological levels, gold objects were imbued with great symbolic weight. They were used to define a social position, to specify an identity or cultural group, to clarify a political role, to indicate an economic status, and to reveal religious allegiances. The multiplicity of roles played by a single object can make deciphering its meaning a complex task, particularly when the artefact was not found in an intact context in which the archaeologists' interpretation could help to lift the veil somewhat on the mystery.

The yellow metal was associated with the work of specialized artisans in a complex chain of operation that necessitated a number of actors

Nose ring, gold and silver
Gold was associated with the Sun, the day, and the masculine. Silver
was associated with the Moon, the night, and the feminine.
Mochica, Costa Norte, Peru, 200 BCE – 600 CE
Museo Arqueológico Rafael Larco Herrera, Inv. ML100866
Diam: 3.1 cm
Photograph: Museo Arqueológico Rafael Larco Herrera, Lima, Peru

from extraction up to goldsmithing. Its most obvious symbolism arose from the shape that the object took and the decorations embellishing its surface. The motifs adorning objects varied enormously and almost always referred to iconography also found on ceramics, textiles, wood, bone, and shells. It must be understood that the symbolic weight of gold objects was complex and their ramifications touched every facet of pre-Columbian societies. Gold objects were goods of prestige, power, reserved for members of the elite, and their magic and religious powers were indisputable. This notion was well rooted in the ideology of the Incas and very probably among their predecessors: "The link between metallurgy and magical transformation has been observed cross-culturally and, in the Andes, the use of metals to represent supernatural forces is a feature not only of the Incas, but of many of their predecessors for over two millennia. In addition to this ideological dimension, metallurgy played a special role in the political economy of the Inca court" (Salazar, 2004: 46). Gold also reigned supreme in the relationship between metallurgy and religious beliefs in Andean civilizations.

At another level, duality seems to have been a fundamental notion in the social organization of Andean groups. The ideological system was profoundly marked by this omnipresent contrast in nature. The most common examples are day and night, Sun and Moon, and fire and water. A number of gold objects with a strong iconographic flavour exhibit this dualist interpretation. The discovery of the Moche royal tombs in Sipán provided confirmation of this duality between gold and silver. Some objects had been made by combining the two metals – two parts brought together to create a single object or composite ensembles. The famous necklace composed of ten gold peanuts and ten silver peanuts is a good example. The gold part of

Silver	Gold
Moon	Sun
Female	Male
Left side	Right side

the necklace rested on the right shoulder of the deceased, and the silver part covered the left shoulder. Hip protectors and *tumi* (ceremonial knives) were also produced in half gold and half silver (Alva and Donnan, 1993: 221–22). It is interesting to note that a gold ingot was placed in the right hand of the deceased and a silver one in his left hand. The association of gold with the right side and silver with the left side is confirmed by discoveries made in intact Sipán tombs. Based on historical data, it is possible to associate gold with the Sun and silver with the Moon. In this regard, the Incas associated gold with the sweat of the Sun and silver with the tears of the Moon (Bray, 1991: 77). A number of historical sources mention the links between gold and the male sex, while silver was associated with the female sex.

Gold was thus opposed to silver, but, as silver was a much rarer metal, it was underrepresented in funerary offerings. Gold and the Sun were therefore preponderant in this Andean duality.

With the centralization of power in the hands of powerful elites in South America, it is generally admitted that gold objects had an inestimable value and that the symbolism surrounding them was undeniably linked to political power, social prestige, and an ideological system. Gold was used to build and confirm cultural and probably ethnic identities. It is possible that some gold objects were associated with the sex of individuals. The value of gold also played a decisive role in trade networks between groups and cultures.

The distribution of gold was very often limited to the ruling class. The fact that objects were discovered in the Moche royal tombs in Sipán (Alva and Donnan, 1993) and in the Sicán tomb in Huaca Loro (Shimada, 1995) confirms that sovereigns took to their tombs enormous quantities of precious objects. Their successors had to ensure that artisans would continue to produce gold objects to accompany them, in turn, to the hereafter.

In Moche funerary practices, the importance of gold is beyond doubt. The quality and quantity of objects made of gold indicated the status of the deceased. Individuals of various social classes were interred with pieces of metal in the mouth or hands. People of the upper classes had pieces of gold; those who were less well off had copper pieces. Although the metals were different, the practice and belief were the same. In addition, neutronic activation analyses conducted on Moche copper objects have revealed that a number of them were in fact *tumbagas*; the gold was no longer visible on the surface, but the chemical content of the pieces indicated that there had been a significant presence of gold (Chapdelaine et al., 2001).

Creativity to Please the Gods

The concentration of gold objects in the hands of a small group accorded "economic" and political power to the owners. The notion of wealth must never be overlooked, even though it did not have the same meaning for each Andean group. These signs of high status were decorated, and their aesthetic value is incontestable. The decoration of certain objects indicates that gold was a medium often used to convey messages other than the preciousness of the metal. Gold objects with complex motifs played a symbolic role in the links between the rulers and the deified forces of nature. The aesthetic aspect therefore seems to be secondary to the sacred meaning of these works.

The diversity of the production of gold objects extended over the entire Andean area. Made by specialized artisans motivated by socio-political and spiritual considerations, many objects came out of Colombian, Ecuadorian, Bolivian, and Chilean workshops. This artistic wealth was of breathtaking dynamism and is a perfect demonstration of the creativity of the artisans working for the great lords, who were considered the representatives of the gods on Earth and did not hesitate to adorn themselves with the sweat of the Sun.

The shapes of objects made of gold responded to aesthetic considerations, and their beauty often resided in the realism of the subjects. The three-dimensional masks were perhaps the best example among the Moche, Sicán, and Chimu. A number of types of objects were used to embellish individuals' appearance, as evidenced by the many bracelets, necklaces, earrings, breastplates, diadems, crowns, plaques, *narigueras* (nose ornaments), and *keros* (drinking cups) found. Large sheets of gold were used to decorate public buildings. The most famous example is the Temple of the Sun, or *corikancha*, located in the centre of the capital of the Inca empire, Cuzco. This temple was vandalized by the Spanish and nothing remains of its grandeur. Designated as the Temple of the Sun by the Spanish, it meant, in Quechua, the Inca language, the house or temple of gold. The witnesses of the time all reported on the extraordinary beauty of the gold-covered walls and the splendour of the enormous garden composed of life-size renditions of plants in gold (Bauer, 2004: 143–52).

A Shell Worth as Much as Gold!

Although all objects made of metal were considered precious – goods with prestige and power – they were not the only luxury goods produced for the elites. According to archaeological data and historical documents, two other categories of objects provided competition. First were fabrics made of wool from the camelidae family, particularly vicuna and, to a lesser extent, alpaca, and a seashell, *Spondylus princeps*, from the warm waters of Ecuador (Cordy-Collins et al., 1999).

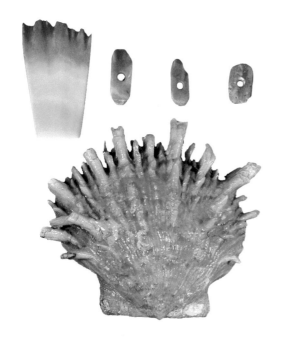

Spondylus princeps
shell called the red gold of the Andes,
a highly prized decorative element.

Photograph: Claude Chapdelaine

Why was this shell so important to Andean peoples? Surprisingly, this trade good had a symbolic value as high as that of gold. *Spondylus princeps* was considered the food of the sea gods. It was accorded a profound significance, often linked to the myth of origin of peoples. Extreme efforts were deployed to ensure its supply. It was sometimes found at more than fifteen metres' depth, and only very experienced divers, specialized collectors, were able to harvest it. The infatuation with this shell developed in the first millennium BCE and never subsided; *Spondylus princeps* continued to be valued over the ages. It was called "red gold," because the red part of the shell was used to make necklaces and jewellery. Even in its natural state, however, this bivalve had an essential importance, and it was a highly valued trade good even before being transformed into luxury objects. Well before the Incas, the Sicáns (Shimada, 1995) and then the Chimu (Pillsbury, 1996) constantly searched for this warm-water shell of the Pacific, gathered at the latitude of Ecuador. In the Andes and on the coast, it has been found in the various elite contexts of ceremonial and urban centres. Its integration into a founding myth, both Andean and coastal, remains a mystery, but its presence alone

confirms that it was a very popular and prestigious object in the early Chavín civilization. *Spondylus princeps* had a place of predilection in the Andean imagination and was coveted by religious elites for several millennia before the Spanish arrived in South America.

Conclusion

For more than 2,000 years, the Andean civilizations accorded great importance to metals, particularly gold. The objects produced were prized in all branches of the society, although they did not have great utility in daily domestic activities. The production of gold and silver objects and, to a lesser extent, copper and bronze objects, was reserved for elites and their immediate supporters. Decorations added to the symbolism and aesthetic of pieces while strengthening their prestigious and magical character.

Gold was at the core of the political economy of the main South American civilizations. It was made into objects conveying prestige, power, ideology, and aesthetics. It had many functions in the worlds of both the living and the dead. In daily life, certain individuals wore gold jewellery that conferred upon them an identity, status, and rank. It was important for the dead to have at their side gold ornaments that would ensure them the same position in the hereafter as they had had during their lifetime. In both worlds, gold objects projected the same messages linked to prestige and power and had a magical and spiritual force.

On the economic level, the acquisition of gold, in all of its forms, was a priority for the elites. In every era, gold was under the control of the wealthy, and its accumulation ensured power. From the beginnings of Andean metallurgy, the metal in its raw state, or more or less cleaned of its impurities, was the medium of exchange. Specialized artisans shaped the identity and originality of each civilization by the forms and decorations that they gave to gold objects. This is why the iconographic richness of gold objects is so great and the multiple functions of these objects refer to a complex fabric of social, economic, and religious motivations. This metal dedicated to the gods, and coveted by their divine representatives on Earth, is, even today, a material that leaves no one indifferent.

Square crown made of gold-plated copper
Prestigious ornament used to convey the rank
or function of the person wearing it.

Viscús, Costa Norte, Peru, 900–200 BCE
Museo Arqueológico Rafael Larco Herrera, Inv. ML100841
H: 12 cm L: 21.5 cm W: 15 cm
Photograph: Museo Arqueológico Rafael Larco Herrera, Lima, Peru

Bibliography

Alva, W., and Donnan, C. B. 1993. *Royal Tombs of Sipán*, pp. 219–27. Los Angeles: Fowler Museum, UCLA.

Bauer, B. S. 2004. *Ancient Cuzco, Heartland of the Inca*. Austin: University of Texas Press.

Bray, W. 1991. "La Metalurgía en el Peru Prehispánico." In S. Purin (ed.), *Los Incas y el Antiguo Peru. 3000 años de Historia*, pp. 58–81. Centro cultural de la Villa de Madrid.

Bruhns, Karen O. 1994. *Ancient South America*. London: Cambridge University Press.

Burger, R. L. 1992. *Chavin and the Origins of Andean Civilization*. London: Thames and Hudson.

Chapdelaine, C., Kennedy, G., and Uceda, S. 2001. "Neutron Activation Analysis of Metal Artifacts from the Moche Site, North Coast of Peru." *Archaeometry*, 43 (3): 373–91.

Cordy-Collins, A., et al. 1999. *Spondylus: Ofranda sagrada y simbolo de paz*. Exhibition catalogue. Lima: Museo Arqueológico Rafel Larco Herrera.

Donnan, C. B. 2003. "Tumbas con entierros en miniaturas: un nuevo tipo funerario Moche." In S. Uceda and E. Mujica (eds.), *Hacia el final del milenio*, vol. 1, pp. 43–78. Actas del Segundo Colloquio sobre la Cultura Moche, Universidad Nacional de Trujillo and Pontificia Universidad Católica del Perú.

Galvez, C., and Briceño, J. 2001. "The Moche in the Chicama Valley." In Joanne Pillsbury (ed.), *Moche: Art and Political Representation in Ancient Peru*, pp. 141–57. Washington DC: National Gallery of Art.

Onuki, Yoshio. 2001. *Una perspectiva del periodo formativo en la sierra norte del Perú*, pp. 103–26. Lima: Historia de la cultura peruana I. Fondo editorial del congreso del Perú.

Pillsbury, Joanne. 1996. "The Thorny Oyster and the Origins of Empire: Implications of Recently Uncovered Spondylus Imagery from Chan." *Latin American Antiquity*, 7 (4): 313–49.

Quilter, J., and Hoopes, J. W. (eds.) 2003. *Gold and Power in Ancient Costa Rica, Panama, and Colombia*. Washington DC: Dumbarton Oaks Publications.

Ramirez, Susan. 1996. *The World Upside Down: Cross-Cultural Contact and Conflict in Sixteenth-Century Peru*. Stanford: Stanford University Press.

Salazar, L. 2004. "Machu Picchu: Mysterious Royal Estate in the Cloud Forest." In R. L. Burger and L. Salazar (eds.), *Machu Picchu, Unveiling the Mystery of the Incas*, pp. 21–47. New Haven: Yale University Press.

Shimada, Izumi. 1995. *Cultura Sicán*. Lima: Fundación del Banco Continental.

Townsend, R. F. 1992. *The Ancient Americas: Art from Sacred Landscapes*. Chicago: Art Institute of Chicago.

Uceda, S. 2001. "Investigations at Huaca de la Luna, Moche Valley: An Example of Moche Religious Architecture." In Joanne Pillsbury (ed.), *Moche: Art and Political Representation in Ancient Peru*, pp. 47–68. Washington DC: National Gallery of Art.

Cóstic Teocuitlatl, Gold of the Aztecs

LOUISE-ISEULT PARADIS

Archaeologist, Professor, Department of Anthropology, Université de Montréal

Translated by Joan Irving

ALTHOUGH CHRISTOPHER COLUMBUS AND the first European explorers were searching for a passageway to the East Indies, they also dreamed of discovering the nations of the El Dorado. They believed they had found them when they began

The Great City of Tenochtitlán, today Mexico City
Fresco on the west all of the National Palace, Mexico City

Diego Rivera (1866–1957), 1945
© Banco de Mexico Trust
National Palace, Mexico City DF, Mexico
Photograph: Schalkwijk/Art Resource, New York

penetrating and conquering the lands today known as Mexico, Colombia and Ecuador, Panama and Costa Rica, and, finally, Peru and Bolivia. These regions had developed a technology and art of metalworking – especially in gold – at least as sophisticated as those in the explorers' countries of origin. The focus of this article is Mesoamerica.[1] Gold appeared in Mesoamerica relatively late in comparison to the regions farther south, namely, around 1300 CE. It nonetheless attained a degree of sophistication that astounded Cortés: "[W]hat can be more wonderful than a barbarous monarch, as he is, should have every object found in his dominions imitated in gold, silver, precious stones, and feathers; the gold and silver being wrought so naturally as not to be surpassed by any smith in the world…?"[2] Such ornaments are familiar to us thanks in particular to the writings of the conquerors and chroniclers who described the region's artisan goldsmiths and techniques, as well as the functions and contexts of gold ornaments. Their accounts deal mainly with the gold of the Aztecs, the last peoples to arrive in the Basin of Mexico and the lords of a vast territory before the arrival of the Spanish. The Spanish of course pillaged and melted down the gold ornaments of the Aztecs or offered them to their sovereign; nevertheless, a few examples remain. With the discovery of the Mixtec Tomb 7 at Monte Albán by the archaeologist Alfonso Caso, the number of gold objects discovered in an archaeological context increased from 32 to 121.[3]

To the Aztecs gold was sacred. In Náhuatl, the word for gold is *teocuitlatl*, meaning divine excrement. The adjective *cóztic* (yellow) carries a specific reference to gold and an association with the Sun god. The modifier *iztac* (white) indicates the silver associated with the Moon goddess. The sacred function of gold is confirmed by the contexts in which it has been discovered and also by the persons bearing the insignia that they have the right to use it.

Origins and History of Goldwork

Although they possessed a few weapons and tools in copper and bronze, the cultures of the New World were still in the Stone Age when the Spanish arrived. Metallurgy developed relatively late in Mesoamerica, after 600 CE. It first appeared in western Mexico: its development seems to be associated with trade between the Pacific coastal regions and the northern coast of South America, where metallurgy had flourished for several centuries. The first metal to be worked was copper. It was not until 1300 that the first gold ornaments appeared, notably in central Mexico and the high plateaus of southern Mexico.

The Players and Their Technology

Fray Bernardino de Sahagún, the veritable ethnographer of the Aztecs, provided the first glimpses of the world of goldsmiths and described their art and technology.[4] Goldsmiths were members of the artisan class who, like precious-stone- and feather-crafters, enjoyed high status in Aztec society. They lived in the Yopico district of the capital Tenochtitlán, where they worshipped Xipe Totec, the god of fertility and rebirth. They used a technology similar to that of the Old World, although this technology had appeared in the Americas much more recently. The Aztecs and the Mixtecs used two main techniques to work gold. The first involved the hammering, in either a cold or heated state, of a sheet of metal that was subsequently decorated using the repoussé technique. The second, described in detail by Sahagún, was the lost-wax technique. After the desired object is shaped on a preform made of clay mixed with ash and dried in the sun, it is covered with a layer of wax and copal resin, and left to harden. The object is then covered by a thin layer of clay mixed with ash and left to dry for two days to prevent cracking. This mould is heated to melt the wax, and molten gold is poured in. When the gold has hardened, the first mould is gently broken and the gold-covered object is polished, before being reheated to give it its final appearance.

---

1. A cultural area geographically situated in southern Mexico, Guatemala, Belize, El Salvador, and the western regions of Honduras, Nicaragua, and Costa Rica. It is made up of similar cultural groups that developed compex civilizations after 1200 BCE.
2. Thatcher, 317–26.
3. Saville, 1920; Caso, 1965: 924.
4. Sahagún, 1950–82, Book 9: 69–77.

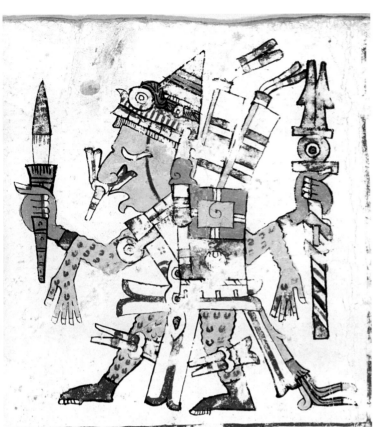

Xipe Totec
He is the god of renewal of nature, agriculture, and beneficial night rain. A powerful divinity venerated with human sacrifices, he was also the patron of goldsmiths.

Codex Borgia
Photograph: Akademische Druck-u.Verlagsanstalt [Adeva], Austria

Quetzacoatl, or the feathered snake god
The god of the morning star, he was considered the inventor of books and the calendar and the one who offered corn to humanity.

Codex Borbonicus
Photograph: Akademische Druck-u.Verlagsanstalt [Adeva], Austria

Contexts and Functions

Gold in its various forms is associated with two main contexts: economic and ceremonial. Gold was one of the forms of tribute that the Aztec state levied on the provinces that it conquered; the *Codex Mendoza* lists six provinces required to pay gold tribute. Among the forms that this tribute took were powdered gold, solid gold bars and disks, a necklace of gold beads, a necklace of gold beads and bells, a diadem, a headband, and a shield. The six provinces were located in the Basin of Mexico, in the state of Guerrero (Tlapan and Quiauhteopan), in the state of Oaxaca (Coayxtlahuaca, Tlachquiauco, Coyolapan), and in Veracruz (Tochtepec). The precious metal was collected from rivers and streams by prospectors who, like those of the Klondike, separated the grains of gold from the sand they had gathered in a strainer or their hands.

Tribute provides information on one way in which the Aztec state obtained gold objects. Gold was probably obtained in other ways as well, namely through long-distance trade with the artisan goldsmiths of Tenochtitlán who transformed it into the ornaments and finery mentioned in relation to numerous celebrations and also found as offerings in funereal contexts. Tomb 7 at Monte Albán provides a good illustration of the riches and beauty of the grave goods of the Mixtec.

Mixtec goldsmiths were great artists, and it is quite probable that the Aztecs learned the art of metalworking through their contacts with them. This tomb must have held the remains of a high dignitary of the Mixtec: among other items found there were 121 gold objects weighing a total of 3,598 kilograms.[5] These were mainly ornaments: pectorals representing deities; rings; necklaces made of beads in the form of jaguar teeth, tortoise

---

5. Caso, 1965: 923–27.

Bell

Mixtec, c. 1500
Museo de las Culturas de Oaxaca, Oaxaca, Mexico
Conaculta-INAH,
Inv. 10-105430, 5.5 cm x 4.5 cm
Photograph: Michel Zabé

shells and bells; tweezers, etcetera. Also found was a piece of gold leaf from the covering of a gourd used to store tobacco and other more fragile items.

My favourite ornament is a pectoral that signifies the universe: its four sections are connected by rings and represent, in turn, Heaven, the Sun, the Moon, and the Earth; hanging from each section is a bell attached to a feather. The whole thing moves and must have jingled delightfully.

In documenting the uses of gold ornaments in rituals and celebrations, the chroniclers, especially Sahagún and Duran,[6] confirmed the divine nature of gold among the Aztecs. During Aztec rituals and celebrations, only the divinities being honoured or the persons – captives or slaves, for the most part – incarnating them had the right to wear

gold ornaments. Thus, the gods Huitzilopochtli, Quetzalcoatl, and Xipe Totec as well as the goddesses Cihuacoatl, Chicomecoatl, and Xochiquetzal, or their incarnations, all wore gold ornaments: pectoral, earings, labret, head ornament edged with a gem set in gold, mitre in gold and precious stones, necklace made up of golden ears of corn, garment embroidered with gold plates. Gold ornaments were also offered in the burial rites of high dignitaries and during celebrations in honour of a god. Of course, members of the elite and the supreme leader also had the right to wear gold ornaments, which accompanied them in death.

Much like cocoa, gold had an economic, ritualistic, and medicinal function among the Aztecs. Those suffering from skin pustules or haemorrhoids ate gold dust or filings. Cecelia Klein has linked the reason for taking gold dust – pustules, or in Náhuatl *nanaoatl* – to the god Nanauatzin, "Our Dear Pustules," who immolated himself during the fifth creation of the universe and became the Sun.[7]

Pectoral portraying the god Xochipilli

Mixtec, 1325–1521
Museo de las Culturas de Oaxaca, Oaxaca, Mexico
Conaculta-INAH, Inv. 10-106163
4.2 cm x 7.3 cm x 1 cm, Photograph: Michel Zabé

6. Sahagún, 1950–82, Book 2; Duran, 1971.
7. Klein, 1993: 4.

Shield-shaped pendant

Mixtec-Aztec, c. 1500
Baluarte de Santiago, Veracruz, Mexico
Conaculta-INAH,
Inv. 10-213084, 10.5 cm x 8.5 cm
Photograph: Michel Zabé

Instituto Nacional de Antropología e Historia   Consejo Nacional para la Cultura y las Artes

Gold and the Spanish Conquest

The circle is complete and we are back where we started, the awe of the Spanish at the riches and gold that they discovered in Mexico. And Motecuhzoma, the supreme leader of the Aztecs, who was both curious and frightened of that handful of bearded men on horseback, and who presented them with numerous gifts before realizing, too late, that Cortés was not the god Quetzalcoatl:

"They laid before them golden streamers, quetzal feather streamers and golden necklaces. And when they had given the gift, they appeared to smile, to rejoice exceedingly, and to take great pleasure. Like monkeys they seized upon the gold. It was as if then they were satisfied, sated, and gladdened. For in truth they thirsted mightily for gold; they stuffed themselves with it, and starved and lusted for it like pigs."[8]

**Anthropomorphic pectoral
portraying Xiuhtecuhtli, the god of fire**
Mixtec, c. 1500, Museo de las Culturas de Oaxaca, Oaxaca, Mexico
Conaculta-INAH, Inv. 10-105429, 6 cm x 4 cm x 1.5 cm
Photograph: Michel Zabé

Instituto Nacional de Antropología e Historia   Consejo Nacional para la Cultura y las Artes

Bibliography

Berdan, Frances F., and Patricia R. Anawalt. 1992. *The Codex Mendoza*, Vol. 4. Los Angeles: University of California Press.

Caso, Alfonso. 1965. "Lapidary work, goldwork, copperwork: Oaxaca." *Handbook of Middle American Indians*, Vol. 3 (2), ed. G. V. Willey, 896–930. Austin: University of Texas Press.

Duran, Fray Diego. 1971. *Book of the Gods and Rites and the Ancient Calendar*. Norman: University of Oklahoma Press.

Galindo Villasana, Ramón. 1997. *Matrícula de tributes*. México D.F.: Secretaria de Hacienda y Credito Público.

Hosler, Dorothy. 1994. *The Sounds and Colors of Power*. Cambridge: The MIT Press.

Klein, Cecelia. 1993. "Teotecuitlatl 'Divine Excrement': The significance of 'holy shit.'" Art Journal (Fall), http://findarticles.com/p/articles/mi_m0425/is_n3_v52/ai_14538980.

Sahagún, Fray Bernardino de. 1950–82. *Florentine Codex: General History of the Things of New Spain*. 13 vols. Trans. A. J. O. Anderson and C. E. Dibble. Salt Lake City: School of American Research and University of Utah.

Saville, Marshall H. 1920. "The Goldsmith Art in Ancient Mexico." *Indian Notes and Monographs* 8. New York: Museum of the American Indian, Heye Foundation.

Thatcher, Oliver J., ed. 1907. *The Library of Original Sources*. Vol. V: 9th to 16th Centuries, 317–26. Milwaukee: University Research Extension Co., http://www.fordham.edu/halsall/mod/1520cortes.html.

8.   Sahagún 1950-82, Forentine Codex, General History of the things of New Spain:Bk.12:31. Translated and edited by A.J.O. Anderson & C.E. Dibble. School of American Research and the University of Utah Press, Santa Fe and Salt Lake City.

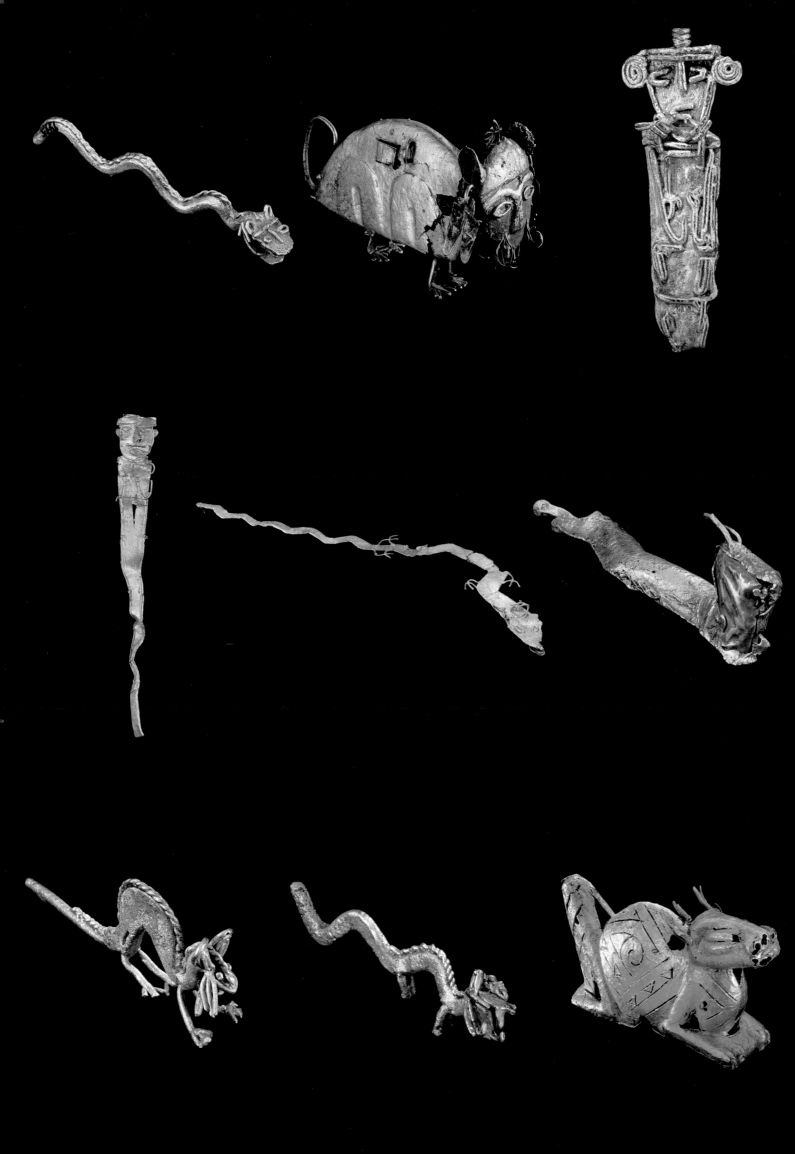

Mutatis mutandis.
The Symbolism of Transmutation
in Muisca Votive Figurines

ROBERTO LLERAS

Archaeologist, Technical Director, Museo del Oro, Bogotá, Colombia

Translated by Joan Irving, from the French translation by Louis Jolicœur

WHEN THE SPANISH CONQUISTADORS arrived in the lands inhabited by the Muisca in the Eastern Cordillera of present-day Colombia, they learned about the existence of a powerful priest or chief called Popón, famous for his ability to travel in air and thus cover great distances, like only a bird could do (Londoño, 2001). Elsewhere, the conquistadors had heard stories about beings that they would have called witches or sorcerers, who changed into jaguars to attack and devour their enemies (Pineda, 2002). These stories about shamans capable of changing into various animals, more common and terrifying than any story the Christians could have imagined, and were always attributed to a demon believed to be the source of these powers (Simón, 1626/1982).

The notion of transmutation is one of the mainstays of the shamanic vision of the cosmos. It seems to be based on the idea that all beings – humans, animals, and plants – have the same constituent elements, as the Uwa of the Eastern Cordillera of Colombia have explained (Osborn, 1990). The possibility that shamans can, as they claim, change into a jaguar, bird, bat, or snake is but one aspect of their special powers. The documents on the mythology of places such as Amazonia report numerous cases of transmutation, showing that this idea is strongly linked to creation myths (Urbina, 1998). Ethnographic studies carried out in contemporary communities reveal that the processes of transmutation generally take place in a state of hallucinatory trance (Pineda, 2002). It is during these rituals, when shamans swallow the hallucinogens known as entheogens and begin the ecstatic flight, that their bodies and identities may change to that of a different being – a great predator, a nocturnal animal, a powerful bird, or a tiny worm, finally possessing all of the abilities of these animals, including that of being able to see the world from such new perspectives (Arhem, 1996).

This is why the traits and attributes of certain animals played such a significant role in the ritual costume of pre-Hispanic shamans, and why they continue to be much in evidence even today (Pineda, 2002). Apart from the costumes, specific objects seem to play an important role in transmutation. Poles, musical instruments, amulets, and figurines are used by shamans during trances as part of the transformation process (Reichel-Dolmatoff, 1988). It is not surprising therefore that these objects, present during the ritual of transmutation, are also an expression of that process. In general, this takes place through a complex iconography that may ultimately be broken down into several main icons that are repeated and multiplied in a variety of periods and places (Reichel-Dolmatoff, 1988).

In the pre-Hispanic metallurgy of the area today known as Colombia, the theme of transmutation appears relatively frequently, although it is manifested differently in each tradition and each style. From the Sierra Nevada de Santa Marta during the Tairona period comes a quadruped form with two extremities, a frog's head and a feline's

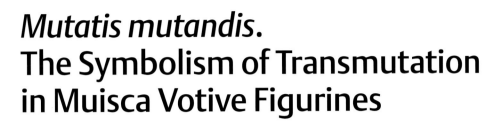

◀ Group of votive figurines depicting transmutation

See also page 106.

head; from the Magdalena Valley, associated with the Tolima style, are found numerous fantastical insects and fish with birds' wings; from the plains of the Caribbean zone come numerous groups of metallurgical works featuring representations of men in the bodies of crustaceans or with frogs' legs; and from the later period of the Quimbayas of the Cauca Valley are found breastplates that mix beings such as men and lizards, to mention only a few of the most remarkable examples (Lleras, 2003). However, the theme encountered most frequently in the pre-Columbian metallurgy of shamanic transmutations in Colombia is that of the man-bird. In his far-reaching study, Reichel-Dolmatoff (1988) identifies numerous expressions of the man-bird icon in the metallurgical iconography.

The Muisca votive figurines produced in the Eastern Cordillera between 600 and 1550 CE constitute one of the most varied and expressive assemblages in terms of iconography. The figurines may be divided into eight groups: women, men, asexual anthropomorphes, scenes, animals, personal objects, domestic objects, and other indeterminate objects. Among the first three groups, some of the figurines are distinguished by the objects they are holding, their costumes, and their postures. Some of the scenes feature one or several persons in specific situations and frameworks. Among the animals, there are different species, especially, snakes and jaguars. The objects include numerous utensils and weapons, as well as various ornaments (Lleras, 1999).

The distribution, chronology, associations, technology, and significance of this unusual collection of votive figurines were thoroughly examined in another article (Lleras, 1999). Here I will limit myself to examining the aspect of transmutation reflected in four series of figurines from different archeological contexts, that is, from different sites and no doubt different eras in the Muisca period. The common denominator linking them is their votive function within each group and the fact that they belong to the same style of metallurgy. The variety of these figurines confirms that the ideas of transmutation upon which such forms of representation are based did not arise in isolation but instead developed in a vast spatiotemporal framework.

The four series are made up of figurines that, through transmutation, link the following: 1) men and felines, 2) men and snakes, 3) snakes and felines, and 4) snakes and deer. Each series is made up of pairs of figurines, each exhibiting its own traits and representing the extremes of transmutation and a variable number of figurines with acquired traits representing intermediary stages in transmutation (for example, the deer with the traits of a snake and the snake with the traits of a deer in the deer-snake transmutation series). The following diagram helps us better understand the nature of each series as well as the links between them.

The existence of multiple series of representations of the process of transmutation between men and animals reveals that the subject of shamanic transmutation is much more complex than the literature in anthropology has indicated. To elucidate upon the symbolism of this grouping, it might be useful to re-examine the animals represented in Muisca mythology as well as the significance of ritual among the Muisca.

When they addressed their people, Muisca chiefs were accompanied by felines – jaguars, or to be exact pumas – representing symbols of authority (Londoño, 2001). The chiefs even had names associated with felines (Pineda, 2002). The Muisca people also had a creation myth with a snake association: it relates how a woman and a child, who had emerged from the Iguaque lagoon, came back as snakes after having put people on Earth (Simón, 1626/1982). Only chiefs could eat the meat of deer (Simón, 1626/1982). The term for a deer in the Muisca language, *Boy chica*, is the same as that used to describe the great civilizer, the lord of agriculture.

The three animals with which man is associated in the Muisca transmutation series have specific qualities that put them into a special

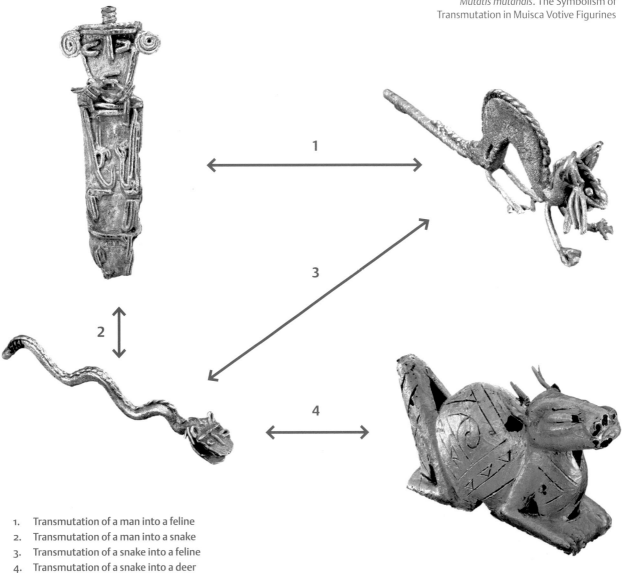

1. Transmutation of a man into a feline
2. Transmutation of a man into a snake
3. Transmutation of a snake into a feline
4. Transmutation of a snake into a deer

category of shamanistic animal likely to become the object of transmutation. But this alone does not explain the existence of these series. Clearly, they represent a specific selection of the transmutations possible in Muisca thinking. On the one hand, several important elements are missing: the man-bird transmutation is not found among the votive figurines in these series; for some reason, it is generally expressed in costume. The bear, the other shamanistic animal whose skin was used by priests (Londoño, 2001), does not appear in the metallurgy. On the other hand, there are some striking inclusions, notably, the snake-feline and snake-deer series expressing transmutations that exclude men.

In contrast to what we might imagine, man is not the central axis of these series of transmutations. Two of the series revolve around the feline.

More important still is the snake, which is central to three series. The deer, on the other hand, is present in only one series. And some potential series are not found: the deer is not transformed into a feline, and man is not transformed into a stag, or vice-versa. If these are the rules of transmutations and the possible links between them, what do they reveal about the mythological and symbolic bases of these processes?

These iconographic series on transmutation probably refer to the permanency of a mythical era when mutations between species were the norm. It was during this era that Bachue became the snake and Bochica, the deer, taught people how to weave robes. This was also a time when man was just one species among many and therefore not at the centre of the cosmos, but rather on the same level as felines and snakes. If we accept this

theory and take into consideration the very nature of the offering, in essence, a plea to maintain the balance of the cosmos (Lleras, 1999), we have a plausible explanation for the appearance of these iconographic series on transmutation in the votive figurines.

These series must have represented offerings intended to re-establish the perfect equilibrium of the beginning of time. By re-creating the multiple transmutations between men and animals through these iconographic series, the metallurgists were evoking the symbolic and religious dimension of the era to which everyone longed to return. Thus did transmutation acquire its true meaning. It was not a process open and accessible to anyone, shaman or not. Transmutation changed that which had to change, enabling a return to the original laws that no longer held sway in the real world. This explained sickness, crop failures, attacks by hostile neighbours. The re-creation of the original transmutations, which might have taken place in the dawn of some cold and forgotten lagoon, opened a door to the ancient world about which there was so much to learn.

Bibliography

Arhem, Kaj. 1996. "Cosmic Food Web. Human-Nature Relatedness in the Northwest Amazon." In *Nature and Society: Anthropological Perspectives*. London and New York: Routledge.

Lleras, Roberto. 1999. "Pre-Hispanic Metallurgy and Votive Offerings in the Eastern Cordillera, Colombia." *BAR, International Series* 778.

Lleras, Roberto. 2007. "La metalurgia prehispánica en el norte de Suramérica: una visión de conjunto." In Roberto Lleras (edo), *Metalurgica en la América Antigua*. Bogotá: FIAN-IFEA.

Londoño, Eduardo. 2001. "El proceso de Ubaqué de 1563: la ultima ceremonia religiosa pública de los muiscas." *Boletín Museo del Oro* 49.

Osborn, Ann. 1990. "Comer y ser comido. Los animales en la tradición oral de los Uwa." *Boletín Museo del Oro* 26.

Pineda, Roberto. 2002. "El laberinto de la identidad. Símbolos de transformación y poder en la orfebrería prehispánica de Colombia." In *Los espíritus, el oro y el chaman*. Salamanca: Fundación La Caixa.

Reichel-Dolmatoff, Gerardo. 1988. *Orfebrería y chamanismo. Un estudio iconográfico del Museo del Oro*. Bogotá: Editorial Colina.

Simón, Fray Pedro. 1626/1982. *Noticias historiales de las conquistas de tierra firme en las Indias Occidentales*. Bogotá: Banco Popular.

Urbina, Fernando. 1998. "Amazônia mítica." In *O mar, eterno retorno. Ourivesaria Pré-Hispanica da Colômbia*. Lisbon: Museu Calouste Gulbenkian.

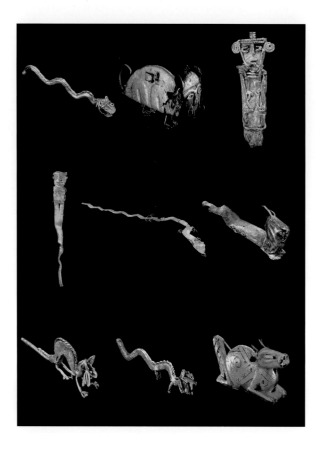

Group of votive figurines depicting transmutation

Muisca, 400–1700 CE
Museo del Oro, Bogotá, Colombia, Inv. O01975
Photograph: Clark Manuel Rodriguez/Museo del Oro

From left to right and top to bottom:

Votive figurine in the form of a snake
Lo : 8 cm, Inv. O01975

Votive figurine in the form of a feline with human face
H: 5.7 cm W 10 cm, Inv. O01115

Votive figurine in the form of a man
H: 5.7 cm W: 2.2 cm, Inv. O02036

Votive figurine in the form of a man with a snake's body
H: 11.3 cm W: 1.6 cm, Inv. O06353

Votive figurine in the form of a snake with arms
L: 17.3 cm, Inv. O23624

Votive figurine in the form of a stag with a snake's body
H: 1.5 cm L: 5 cm, Inv. O02314

Votive figurine in the form of a feline
H: 2.3 cm L: 5.8 cm, Inv. O06303

Votive figurine in the form of a feline with a snake's body
Inv. O33055

Votive figurine in the form of a stag
L: 4.5 cm, Inv. O33078

A Mochica
Ritual Garment

SANTIAGO UCEDA CASTILLO

Archeologist, Director of the Huaca de la Luna archeological project, Universidad Nacional de Trujillo

Translated by Käthe Roth, from the French translation by Louis Jolicœur

Details of the frescos in the interior court of the Temple of the Moon

Photographs: Santiago Uceda Castillo, 2007

THE SITE OF THE Mochica temples of the Sun and the Moon, an archaeological complex covering about 60 hectares, is located on the outskirts of the modern-day city of Trujillo in Peru. It is made up of two monuments in the form of graded pyramids, with a dense urban zone spread out between them. The goal of the archaeological excavations, which began in 1991, is to examine the sequence of occupation, the structure, and the function of the various components of the site.

The work on the Temple of the Moon has revealed five superimposed buildings representing almost six centuries of Mochica occupation. These layers of construction were part of a process of power renewal in Mochica society (Uceda, 1997). The relief work on the façade, the interior spaces of the structure, and the objects discovered there, clearly indicate that a variety of ceremonies took place at this temple, including the ritual of human sacrifice (Uceda and Tufinio, 2003). These ceremonies, as well as the temple burials, served as mechanisms for

Other details of the frescos
in the Temple of the Moon

Photographs: Santiago Uceda Castillo, 2007

Mochica ritual garment
in the form of a feline skin,
discovered in the Huaca
de la Luna site in Peru ▸

Universidad Nacional de Trujillo
Peru, Projet Huaca de la Luna
Inv. PHLL-56 INC-03
H: 67 cm W: 31 cm D: 11 cm
Photograph: Steve Bourget

The materials placed on the fabric inside the reed basket were retrieved layer by layer, as during any laboratory examination, but no order was observed that might indicate that the materials were part of a more complex series. The only exception was the piece of fabric decorated with plates made of gold or other materials that represents the skin and face (in relief) of a feline.

When this piece was discovered, it was complete except for some parts of the feline's body, and the head had been flattened by the weight of the fill. Not to go into too much technical detail, note simply that the object is made up of four main elements, the main one being an animal skin cut to indicate the general form of the object (the stretched skin of a feline). The skin is covered on each side by fabric onto which are sewn metallic disks of various shapes that contribute to the overall appeal and motif of this fine garment. The most visible pieces of metal are the four paws, which were no doubt cast using the repoussé technique. The spots on the feline's fur are made from metal plates in the form of overhanging sequins, and black and yellow feathers inserted where there are no sequins. The tip of the tail is adorned with a metal disk painted with cinnabar.

A head, made from tree resin, was added to the initial leather and fabric structure. The animal's

legitimizing political power and the role of the elites at the centre of Mochica society.

The Mochica ritual garment presented here was discovered during the excavations of the fill covering Building D on Platform I of the Temple of the Moon. Six elements were found on top of and around Tomb 18; archaeologists have named these "items" (Tufinio, 2004). Item 4[1] is, in our opinion, a reed basket linked to the tomb or an offering related to the burial ritual practised at the temple (Building D in the construction sequence of this temple; Uceda and Canziani, 1998).

---

1.   Item 4 was lofelineed 90 cm above the ground of Building D and 60 cm from MA1. The most important item was found between quadrilaterals 1CQ and 1CR. It is a basket measuring 60 x 46 x 45 cm and made of strips of reed stems attached by cotton cord. Inside, on a woven cotton fabric decorated with black and yellow feathers in the form of wreaths, various laminated gold objects were found, as well as a piece of fabric decorated with feathers and a garment shaped like the skin of a feline.

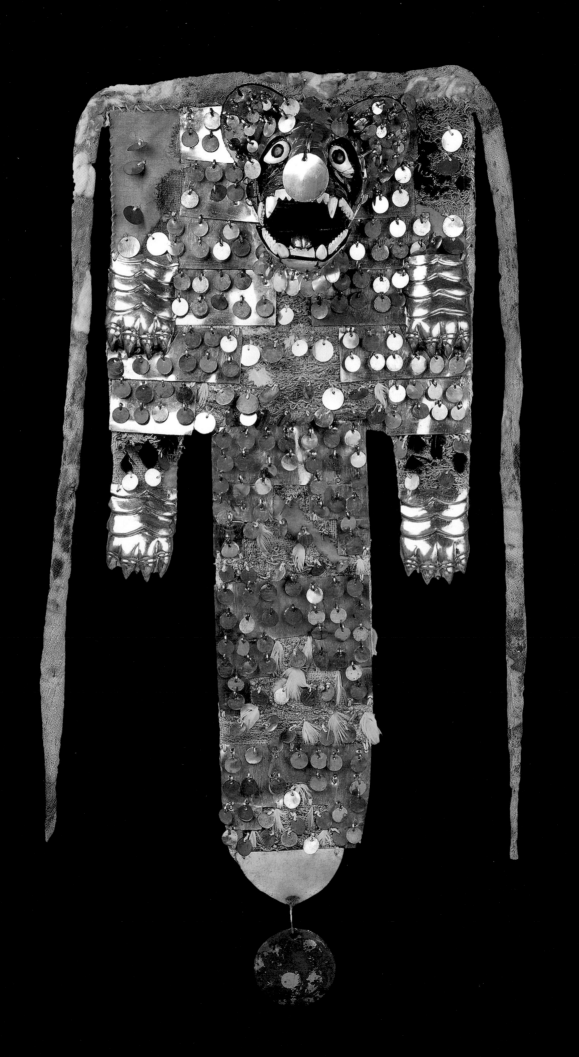

Chacchadores
Illustration of the ritual consumption of coca.

Linden Museum, Stuttgart
Photograph: Christopher B. Donnan

mouth is open and reveals teeth made from *conus* shell. The tongue and palate are made from one piece of hollowed copper. The two eyes are also made with *conus* shell, and the pupils are formed by two black stones. In addition to these elements, there is a ring on the animal's muzzle and a chin-strap made from cut and cast gold plates. Sewn along the top of the piece are fabric ties filled with cotton; these were used to tie the garment on to the wearer's back.

The Ceremonial Garment
in Mochica Rituals of Human Sacrifice

What was the function of this ceremonial garment in Mochica society? A scene in the iconographic art of the Mochica shows mythical figures and priests putting on and receiving this garment during a ritual evoking the chewing of coca leaves. In another article, we suggested that this scene is closely linked to the theme of human sacrifice as well as to ritual combat (Uceda, 2004).

Recent iconographic studies on the theme of human sacrifice have revealed a sequence of images beginning with combat and continuing with the taking of a prisoner, the parading of the nude prisoner, and his preparation and execution. The prisoner's throat is cut and his blood is gathered in cups bearing representations of figures dressed as divine beings. Such sequences, with their variations and different number of figures, describe the ritual of human sacrifice among the Mochica

as well as various activities whose meaning is still being discussed.

Rafael Larco was the first to set forth the theory that the sacrificed individuals were members of groups conquered by the Mochica. Some archaeologists agreed with this approach and went on to characterize the Mochica as an expansionist military state (Moseley, 1992; Lumbreras, 1969; etc.). More recently, Krzysztof Makowski (1997) as well as Luis Jaime Castillo and Ulla Holmquist (2000) have suggested that the sacrificed individuals were instead linked to a ritual battle. For its part, the Department of Anatomy and Physiology of the Saga Medical School in Japan conduced mitochondrial DNA tests on a group of skeletons from sacrificed individuals and priests buried in the Temple of the Moon, as well as on others buried in the residences located between it and the Temple of the Sun. These studies showed that, with one exception, the individuals all belonged to the same ethnic group and were related by blood through more than five generations (Shimada et al., 2002, 2006). The present study, however, reveals a series of consequences linked not only to the origin of these sacrifices but also to the very structure of Mochica society.

First, acknowledging that the persons being sacrificed came from the same Mochica group, these are therefore scenes of ritual battle between different groups and sub-groups who shared a territorial, social, and political reality, such as are found in numerous early and even present-day Andean societies. Second, this sacrifice was not a demonstration of power, as in the case of societies that took prisoners from different but neighbouring communities. It was, instead, an offering by the community to its gods. This offering acted as both a mode of communication and a means to consolidate relations within the group and towards its gods and leaders. Third, as these individuals were blood relations going back five generations, other social and political consequences were no doubt involved.

The ritual battle did not serve solely to "choose" the sacrificial victim; it must also have served as a catalyst to the conflicts inherent in a highly stratified society such as that of the Mochica. This symbolic garment is identical to those in the scene evoking the chewing of coca, in which a divinity (the Divin Twin, according to Makowski Hanula) wears the same symbol on his shoulder.

In other scenes a warrior wears this symbol on his back; it is passed to the victorious warrior by the hand of the gods, which may signify that these symbols were associated with acts linked to the ritual of human sacrifice and thus became a mechanism for social advancement aimed at alleviating social conflict (Uceda, 2004).

This discovery is significant in that it enables us to reinterpret several narrative sequences from Mochica iconography (Uceda, 2004) and therefore to show that these iconographic representations evoke rituals that were fundamental to the power structure of Mochica society.

Warriors wearing ritual costumes in the form of feline skins

Ceramic vase
Cleveland Museum of Art
Photograph: Christopher B. Donnan

Bibliography

Castillo, Luis Jaime, and Ulla Holmquist. 2000. "La ceremonia del sacrificio mochica, en el Museo Arqueológico Rafael Larco Herrera." *Revista de Arqueología* 11 (232): 54–61, Madrid.

Lumbreras, Luis Guillermo. 1969. *De los pueblos, las culturas y las artes del antiguo Perú*. Lima: Moncloa-Campodónico, Editores Asociados.

Makowski Hanula, Krzysztof. 1997. "La guerra ritual." *Perú El Dorado* 9: 62–71. Lima: PromPerú.

Moseley, Michael E. 1992. *Incas and Their Ancestors. The Archaeology of Peru*. London: Thames and Hudson Ltd.

Shimada, Izumi, Ken-Ichi Shinoda, Steve Bourget, Walter Alva, and Santiago Uceda. 2002. "MtDNA Analysis of Mochica and Sican Populations of Pre-Hispanic Peru." In *Biomolecular Archeaology: Genetic Approaches to the Past*, ed. David Reed, 61–92. Carbondale: Center for Archaeological Investigations, Southern Illinois University.

Shimada, Izumi, Ken-Ichi Shinoda, Steve Bourget, Walter Alva, and Santiago Uceda. 2006. "Estudios arqueogenéticos de las poblaciones prehispánicas mochica y sicán." *Arqueología y Sociedad* 17: 223–54. Lima: Museo de Arqueología y Antropología and Centro Cultural de San Marcos, Universidad Nacional Mayor de San Marcos.

Tufinio, Moisés. 2004. "Excavaciones en la Unidad 12a (ampliación norte), Plataforma I, Huaca de la Luna." In *Investigaciones en la Huaca de la Luna 1998–99*, ed. S. Uceda, E. Mujica, and R. Morales, 21–39. Trujillo: Facultad de Ciencias Sociales de la Universidad Nacional de Trujillo.

Uceda C., Santiago. 1997. "Le pouvoir et la mort dans la société Moche." In *À l'ombre du Cerro Blanco, nouvelles découvertes sur la culture Moche, côte nord du Pérou*, ed. C. Chapdelaine. *Les Cahiers d'anthropologie* 1: 101–16. Département d'anthropologie, Université de Montréal.

Uceda C., Santiago. 2004. "Los sacerdotes del Arco Bicéfalo? Tumbas y ajuares hallados en Huaca de la Luna y su relación con los rituales Moche." In *Informe Técnico 2003*, Proyecto Arqueológico Huaca de la Luna, ed. S. Uceda and R. Morales, 237–59. Trujillo: Facultad de Ciencias Sociales de la Universidad Nacional de La Libertad – Trujillo.

Uceda, Santiago and José Canziani. 1998. "Análisis de la secuencia arquitectónica y nuevas perspectivas de investigación en Huaca de la Luna." In *Investigaciones en la Huaca de la Luna 1996*, ed. S. Uceda, E. Mujica, and R. Morales, 139–58. Trujillo: Facultad de Ciencias Sociales de la Universidad Nacional de La Libertad – Trujillo.

Uceda, Santiago and Moisés Tufinio. 2003. "El complejo arquitectónico religioso Moche de Huaca de la Luna? Una aproximación a su dinámica ocupacional." In *Moche? hacia el final del milenio*, ed. Santiago Uceda and Elías Mujica, 179–228. Actas del Segundo Coloquio sobre la Cultura Moche (Trujillo, 1 to 7 August 1999), Vol. 2. Lima: Universidad Nacional de Trujillo y Pontificia Universidad Felineólica del Perú.

The Prospector
by Robert Service

It was my dream that made it good, my dream that made me go
To lands of dread and death disprized of man;
But oh, I've known a glory that their hearts will never know,
When I picked the first big nugget from my pan.
It's still my dream, my dauntless dream, that drives me forth once more
To seek and starve and suffer in the Vast;
That heaps my heart with eager hope, that glimmers on before –
My dream that will uplift me to the last.

Perhaps I am stark crazy, but there's none of you too sane;
It's just a little matter of degree.
My hobby is to hunt out gold; it's fortressed in my brain;
It's life and love and wife and home to me.
And I'll strike it, yes, I'll strike it; I've a hunch I cannot fail;
I've a vision, I've a prompting, I've a call;
I hear the hoarse stampeding of an army on my trail,
To the last, the greatest gold camp of them all.

CHAPTER **5**

THE GOLD RUSHES

The California Gold Rush: El Dorado Captured, California Redeemed
PAUL-CHRISTIAAN KLIEGER, Ph.D., anthropologist, chief curator, History Division
Oakland Museum (California)

The Klondike Gold Rush, 1897–98
MICHAEL GATES, museologist and archaeologist, Yukon National Historical Park

Gold Rushes in Quebec
JOSÉ LOPEZ ARELLANO, anthropologist, head of the exhibition's Research Committee,
Musée de la civilisation

Fool's Gold
YVES CHRÉTIEN, HÉLÈNE CÔTÉ, RICHARD FISET AND GILLES SAMSON, archaeologists
Commission de la capitale nationale du Québec

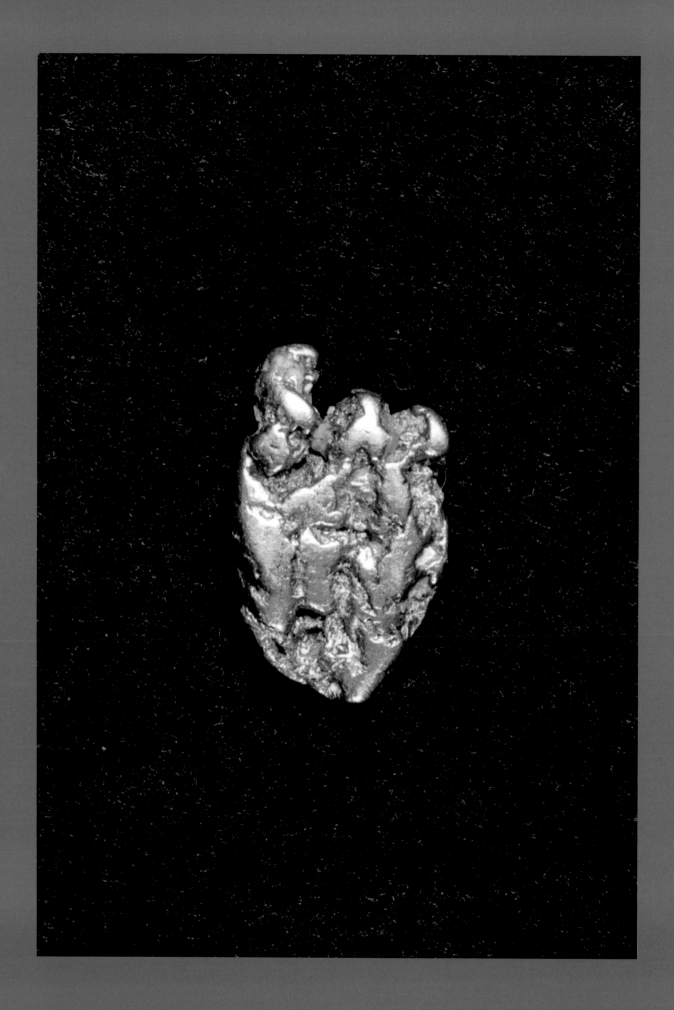

The California Gold Rush: El Dorado Captured, California Redeemed

PAUL-CHRISTIAAN KLIEGER

Anthropologist, Chief Curator, History Division, Oakland Museum (California)

CALIFORNIA WAS PART OF the last Spanish out-post established in the New World. On January 24, 1848, with only nine Mexican sunsets left for California, Peter Wimmer and James Marshall plucked a shiny nugget out of the American River while helping John Sutter build his sawmill at Coloma near modern Sacramento.

Within an instant, the future of the sleepy Mexican province of California was forever changed. The news of gold was soon spread by brash newspaperman Sam Brannan, a prominent citizen of the little town of San Francisco, who boldly announced it to the world. The California Gold Rush was on. Within months, tens of thousands of "Argonauts" from around the world took the long trip to California, settling this isolated region in a blink of the eye. Fate seemed to have smiled on the United States, for, after three hundred years of Spanish and Mexican rule, a rule that was founded on a quest for riches, and especially gold, California seemed to have redeemed itself in the eyes and by the hands of Yankees, not Latin Americans.

Nine days after the discovery, with Mexico defeated in the Mexican-American War, California was annexed to the United States under the Treaty of Guadalupe-Hidalgo. The immense region

moved directly to statehood by 1850 as the "Golden State." The great city of San Francisco was founded on the Gold Rush, jumping from a little village of 800 to a city of 100,000 in a few years. It quickly became the premier city of the West Coast of the Americas. All this because of the discovery of gold . . . or so it would seem.

The story of the California Gold Rush and the taming of the Old West is an entrenched American archetype, an ordained conclusion to the Manifest Destiny of the republic, all centred on that winter's day at a lonely sawmill in 1848. The gold seekers who underwent the long journey to California between 1848 and 1854 represented the greatest mass migration in American history to date, with perhaps 300,000 individuals arriving within a few short years. California has subsequently mushroomed to be the country's most populous and wealthiest state, with 36 million people. It would be the world's fifth-largest economy if compared with nation-states.

If this irony seems a little too stiff, it is. The "discovery" of gold in California in 1848 parallels Columbus's "discovery" of the New World. Just as people were already in the Americas in 1492, gold was already known in California before 1848. The apparently seamless transition from renegade Mexican province to valued Yankee state has been hammered into the dominant historical narrative by Marshall's gilded discovery. The telling of the Gold Rush story in the popular U.S. press at the time provides clear examples in the discourse of strained ethnic relations, culminating with the

◄ First gold nugget found in California
by Peter Wimmer and James Marshall, 1848

Photograph: Courtesy of the Bancroft Library
University of California, Berkeley

Gold jewel casket, 1869–78
The inside of the cover of this chest, made of California gold and quartz, is a relief presenting a scene depicting the settlement of the West. The newly completed transcontinental railway crosses through fertile plains in which Native Americans are chasing bison.

A. Andrews ~ goldsmith
H: 17.5 cm W: 22.5 cm D: 17,5 cm
Collection of the Oakland Museum of California,
Gift of Kenneth Brown

Aztec Gold

The idea that Amerindians of California "had no use" for the little gold nuggets, which they undoubtedly saw in creek beds along the western slopes of the Sierra Nevada Range, is an ethnocentrism inherited from the patriarchal Anglo historical paradigm that succeeded at California statehood. The great empires of Native America, such as the Aztec, Maya, and Inca, certainly utilized gold, especially for ceremonial and personal decoration. It was considered the "excrement of the gods"[2] by the Aztecs and was traded for with other luxury goods such as quetzal feathers and turquoise. The social impact of the great Valley of Mexico imperial Native states must have been felt in modern-day Sonora and up to California through trade, linguistic, and socio-cultural affinities. As in all empires, riches flowed from the frontier to the centre, and the deserts of eastern California were near the frontier of the Native Mexican state at its greatest extent. Could California gold have been traded southward? Perhaps there are oral traditions in Native Californian memory regarding the pre-Contact use and trade of gold.

supposed Yankee "victory" over the Hispanic colonialists and the American Native peoples of the frontier. As Rohrbough suggests, "These national heroes (and some heroines) would make California an American land, replace Mexican Catholicism with American Protestantism, supplant the Spanish language with English, and supersede Mexican culture with American values and institutions."[1] In a reversal of the alchemical mandate, gold became the medium rather than the product of a wondrous social transmutation.

1.   Malcolm Rohrbough, "No Boy's Play," in Kevin Starr and Richard J. Orsi (eds.), *Rooted in the Barbarous Soil* (Berkeley: University of California Press, 2000), p. 30.
2.   The Aztec Empire: Noble Life and Everyday Life. Arts Curriculum Online. Guggenheim Museum. [http://www.guggenheim.org/artscurriculum/lessons/aztec_L6.php]

Spanish and Mexican Gold Mining

Despite this speculation, the first written, although sketchy, accounts of gold in California date to the 1790s and the mysterious Lost Padre mine. This was an alleged discovery made in the San Emidio mountain range south of Fort Tejon and north of the present-day Castaic reservoir, all about 30 miles north of modern Los Angeles. In 1862, an old prospector stumbled into Fort Tejon with a heavy sack of gold that he said he had gathered south of the fort. A local Amerindian named Tucoya related that long before he was born some white men had come to his village. Each wore long robes with a rope tied around his waist. Building a smelter at San Emidio Canyon, they enlisted Tucoya's father and other Native men to help them dig out the rich tailings of gold and smelt them into bullion. Each spring, up to 100 packhorses and 30 to 40 men would transport the gold southward and across the Colorado River to an unknown destination. This continued for years, until Tucoya was old enough to see for himself. The priest he encountered at the mine warned him to keep the secret. Eventually a band of Paiute attacked the village of San Emidio, killing the priests and most of the Indians. Tucoya was the only one left alive by 1862. Then he attempted to show his colleagues the location of the mine, guiding them to within a few miles of the diggings. But he could go no further; he could not violate the secret. The secret died with him, and no one has since found the lost mine.[3]

Another researcher reveals that this area was mined by the Jesuits long before the Franciscan fathers established their missions under Father Junípero Serra. The Jesuits had established their original missions in Baja California (1695). In 1767, the Spanish king, Carlos III, banished the Society of Jesus from the imperial lands throughout the empire. Dominicans replaced Jesuits in Baja, while Franciscans were instructed to establish missions in Alta California.[4] The Jesuits escaped to Germany or Russia, leaving, according to

Sparacino, their last store of Lost Padres gold safely in the mine, which was then carefully disguised.[5] The *Pasadena Union* of 29 October 1887 indicates that the Native peoples working the mine in those early days led an insurrection against the padres and destroyed all visible traces of the mine.[6]

Some of the other mines in the region were subsequently worked by Franciscans. In 1820, a party under the leadership of Santiago Feliciano explored the Castaic region and travelled up

Hasley Canyon, where they discovered gold in a camp they named San Feliciano.[7] From 1834 to 1838, the adjacent San Francisquito and Placerita Caceta placers were worked by Franciscan priests from the San Fernando and San Bueno Ventura missions. In 1832, Ewing Young found a gold-smelting oven in San Emigdio Canyon.[8]

The classic tale of Mexican gold in California is the Rancho San Francisco (Newhall Ranch) find of 1842 near modern Santa Clarita in Placerita Canyon. It was not broadly publicized – common sense, no doubt, but perhaps reflecting the entrenched mythology of the Lost Padres Mine.

3. Virginia Wegis, "Legend of the Los Padres Gold Mine," *Daily Midway Driller* July 25, 1979.
4. Kent Lightfoot, *Indians, Missionaries, and Merchants* (Berkeley: UC Press, 2005), pp. 52–53.
5. Peter Sparacino, [http://www.a1pro.net/~kb6dj/UpperC.htm].
6. Bob Kerstein, History of the famous Monte Cristo Gold Mine. [www.encyberpedia.com/gold.htm]
7. Charles Outland (1986) in Leon Worden, New Study Will Nag SCV Historians. August 14, 1996. www.scvhistory.com/scvhistory/signal/worden/lwo81496.htm
8. Ibid.

Gold revolver

Colt, 31-calibre
J. W. Tucker & Co., goldsmith, 1849
San Francisco, California
Collection of Greg Martin
Photograph: Douglas Sandberg

Governor Juan B. Alvarado had granted a rancho here to Mexican Lt. Antonio del Valle. Stocked with cattle, sheep, and horses, the rancho was head-quartered at the Asistencia de San Francisco Javier at present-day Castaic Junction. Jose Francisco de Gracia Lopez, an uncle of Don Antonio's second wife, leased some of this land from his in-law and ran his own cattle. On March 9, 1842, noontime on the day of his fortieth birthday, Lopez was deep in Cañon de los Encinos (Live Oak Canyon), picking a spot under an ancient oak tree for lunch and a siesta. He fell asleep and dreamed that he was float-ing on a pool of gold. After his nap, Lopez dug up some wild onions with his knife and was surprised to discover gold clinging to their roots.[9]

Lopez and his associates scavenged the river-banks and came up with more of the yellow metal. They took their booty to Los Angeles and sent word to Mexico City. Assayed by the Philadelphia Mint, the gold tested out at .926 fine. A petition

for a claim was filed with the Mexican governor Alvarado, becoming the first documented gold dis-covery in the region. News of the strike was even published in the New York *Observer* on October 1, 1842: "They have at last discovered gold, not far from San Fernando, and gather pieces of the size of an eighth of a dollar."[10] But the thousands of Yankees did not budge – California was a part of Mexico. Hundreds of prospectors from Los Angeles and Sonora, however, did flock to Cañon de los Encinos, which was renamed Placerita Canyon. "Placer," a word of of Spanish origin, means sur-face deposits of sand or gravel containing gold.

From 1842 to 1847, the miners culled some 1,300 pounds of gold from Placerita. Don Antonio del Valle's son Ygnacio became head of the mining district centred there. In 1845, northern Californian settlers John Sutter and John Bidwell, with an army of mercenaries, were enlisted by the unpopular governor, Manuel Micheltorena, to serve in fight-ing the insurrectionist former governor Alvarado in southern California. After the "comic opera" Battle of Cahuenga in the San Fernando Valley, north of modern Hollywood (where Universal Studios now sits), Sutter and Bidwell were thrown into jail by victorious Alvarado. Governor Micheltorena was deposed and fled to Sonora, while Sutter and Bidwell were released. Leaving the jail near Mission San Fernando, Bidwell went north through Placerita Canyon, where he ob-served the gold mining operations. Undoubtedly, both Sutter and Bidwell were intrigued about the possibility of striking gold in their own lands in great Central Valley of California. When the Treaty of Guadalupe Hidalgo of 1848 turned California over to the United States, and many of the southern California Sonoran miners went home, the story of the southern California "Gold Rush" faded into oblivion. Today, "the Oak of the Golden Dream" is an obscure California historic landmark.

9.   Leon Worden, "California's Real First Gold," *COINage magazine*, October 2005. www.scvhistory.com/scvhistory/signal/coins/worden-coinage1005.htm.
10. Sydney Morse (brother of Samuel). Lopez Gold Discovery in NYC Newspaper, 1842. Santa Clarita Valley History in Pictures. www.scvhistory.com/scvhistory/lw2181. htm

Belt buckle

Belt buckle decorated with a portrayal of Minerva with a bear, iconographic elements borrowed from the official seal of the State of California

Manufactured by California Jewelry Company, patented by William Cummings, c. 1868
Metropolitan Museum of Art, purchase, gift of Susan and Jon Rotenstreich, 2000, Inv. 2000.571
Photograph: © Metropolitan Museum of Art

Manifest Destiny

It seemed as if Divine Providence herself scribed California's chapter of Manifest Destiny – a presumption of President James Polk and his Democratic party – that the United States was justified in expanding its realm from the Atlantic to the Pacific. The phrase was first used by journalist John O'Sullivan in his writings urging the annexation of the Republic of Texas, but antecedents of Manifest Destiny extended back to the Jeffersonian and Jacksonian ideals of an "empire of Liberty." The Sutter gold finding along the American River

in 1848 became an expedient in the nation-building enterprise of the United States. The notion that the Anglo-Saxons of the United States would "naturally" prevail throughout the continent, especially after the precipitous decay of the Spanish and Mexican realms, was charged with racial, religious, and political extremism. The importance of the Sacramento Gold Rush story, as opposed to the southern gold strikes in the region, clearly illustrates U.S. nationalistic bravura. The discovery of gold by and for American settlers symbolically increased the value of the region to Washington. At the rather convenient conclusion of the Mexican-American War in 1848, the direction was clear. To avoid squabbling over whether California would be a slave-holding or free state, and issue starting to seriously divide the country, the California region was rapidly pushed to statehood in 1850.

Whereas the Spanish padres, the Amerindians, and the Mexicans kept their gold mining activities in California strictly to themselves, it was the adjacent merchants in San Francisco, such as Samuel Brannan, who had most to gain by spreading gold fever to as many prospective customers as possible. Even Sutter had tried to keep Marshall's findings secret. News of the northern findings was published in the *New York Herald* on August 19, 1848. On December 5, President Polk himself announced the discovery in an address to Congress. The effect of this gilding of California, a trophy of war, was immediate and profound. Whereas only about 2,000 Mexicans had joined the Placerita

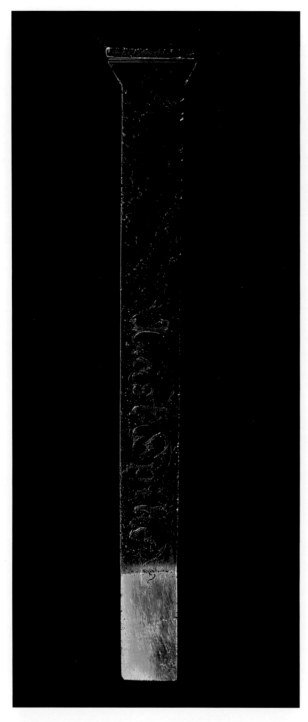

Gold nail marking the completion of construction of the railway link between San Francisco and Los Angeles, September 5, 1876

California Historical Society, Crocker Collection, Inv. X92-3-75
Photograph: California Historical Society

squatters and his own workers left his employment to work the tailings. He was ruined. San Francisco's harbour was soon crammed with abandoned ships as miners left for the goldfields, but its merchants prospered – selling picks, shovels, pans, and food-stuffs to outfit the Argonauts.

By 1854, most of the surface gold in California had been gathered, leaving others with the need to invest in heavy equipment to blast, crush, and smelt gold-bearing rock or blast away the soil matrix with high-pressure water hoses. Wealth continued to flow into places like San Francisco. In 1859, the boom was sustained with the discovery of a large deposit of silver ore on the eastern slopes of the Sierra Nevada near the gold fields. Much of the wealth of this "Comstock Lode" went to build San Francisco, worth approximately $500 billion to $600 billion in today's dollars.

The era of the Gold Rush could be said to have ended at the other great symbol of American Manifest Destiny, the completion of the first trans-continental railroad in 1869. In one of the United States' most iconic images, the locomotives of the Central Pacific from the West and Union Pacific Railroad from the East met, cowcatcher to cow-catcher, on May 10 at Promontory Point, Utah.[11] California governor Leland Stanford, president of the Central Pacific Railroad, hammered down the last spike, suitably made of solid 18-karat gold. California was officially bound to the United States by this obviously gendered impalement. Gold had been the motivation to open the Frontier; gold was there to symbolize its end.[12]

In sum, the California Gold Rush "spoke to American values at mid-century: the democratic belief that wealth would be available to all, that success would reward hard work, and that the largest portion of riches would go to those most moral and worthy by the standards of the day."[13] And these "standards" involved the replacement of Spanish and Mexican Catholicism with Yankee Protestantism, Spanish with English, and fate with individual enterprise, all as ordained by Divine Provenance.

diggings in Southern California, over 300,000 would be travelling to Sutter's lands in the north, by land and sea. Sutter himself was overrun by land

---

11. Andrew Russell photograph, original glassplate negative at Oakland Museum of California.

12. There were seven additional transcontinental railroads built in the United States and Canada after 1869, each replicating the Promontory event with last spike ceremonies. The curtain closed on the era at the completion of the San Diego & Arizona Eastern Railroad, built at great expense by John Spreckels. When it came time to lay the last rail in 1919, he of course used a golden spike.

13. Rohrbough, "No Boy's Play," p. 28.

The Klondike Gold Rush, 1897–98

MICHAEL GATES

Museologist, archeologist, and former cultural resource manager for the Yukon Field Unit of Parks Canada

The Chilkoot Pass, Alaska

Library and Archives Canada, Inv. C4490
Photograph: Library and Archives Canada

THE SEARCH FOR GOLD in the Yukon started in 1874 with the arrival of the first of many prospectors. Among them, Arthur Harper, Al Mayo, and Jack McQuesten were the leaders in the quest for gold for the next quarter-century. They encouraged, promoted, and then supplied the prospecting brotherhood that developed slowly before the gold rush. At first a trickle, then a steadily increasing stream of hopeful prospectors entered the Yukon River basin, spurred on by the increasingly promising reports of gold on the bars of the Yukon and then its tributaries: the Stewart River (1885), the Fortymile River (1886), the Sixtymile River (1891), and finally Birch Creek, near Circle City, Alaska (1892).

The Yukon presented the harshest conditions of any of the gold rushes: long, dark winters with

A merchant and his customers in front of the Northwest Trading and Commission Co. store, Dawson City

temperatures often reaching –50°C, interrupted by short summers of perpetual daylight. With the rivers frozen and the mountain passes clogged with snow, thousands of kilometres from the large urban centres of Canada and the United States, the early prospectors faced months of total isolation and deprivation. The ground in which they mucked for gold was perpetually frozen, harder than granite. In order to unlock nature's vault, they had to master techniques of thawing the ground that were later employed in the Klondike.

The early prospectors relied heavily upon the First Nations people, who had lived in this land for thousands of years. They used their trails, learned their strategies for survival, and married their women. The traders exchanged goods for their furs.

Early mining was supported and encouraged by the grubstake system, whereby traders such as Harper, McQuesten, and Joseph Ladue extended long credit to prospectors so that they could continue the quest for the next rich find. When new discoveries were made, the miner's code ensured that this information was widely shared. The trading posts were the hubs at which information was exchanged; this was encouraged by the operators of the posts, as any significant discoveries by the miners whom they had supplied would attract more business when others flocked to the new

discovery. Thus it was that when Joseph Ladue, at his trading post on the Yukon River, 150 kilometres above Fortymile, started directing miners into the Klondike Valley in 1896, he was quick to capitalize on the discovery of gold on Rabbit Creek. While others staked claims for gold, Ladue staked out 160 acres at the mouth of the Klondike River, called it Dawson City, and made a fortune selling lots and the lumber to build on them.

The discovery in August of 1896 on the Klondike tributary that was quickly renamed Bonanza Creek was the one that dreams were made of. Within months, word of the incredible wealth emerging from the mineshafts of Bonanza, Eldorado, and dozens of other streams started to trickle to the outside world. When tattered and weathered sourdoughs disembarked laden with

Looking at the Golden Stairs
from the scales, Chilkoot Pass
Everyone who wished to reach the prospecting sites had to transport the materials needed for their survival there for one year – one ton of materials.

gold dust from the steamship *Portland* at a dock in Seattle on July 17, 1897, the news sparked an unprecedented stampede. What followed was a mass odyssey of such proportions that it left the memory of the adventure more indelibly etched in the psyche of a generation than it left gold in the pockets of those who embarked upon it.

It was an epic journey during which numerous challenges had to be met and countless obstacles overcome. First, there was the voyage north along the Pacific coast in bad weather aboard leaky, overloaded ships of every description, followed by arrival at the two coastal Alaskan ports of Skagway and Dyea, the former a lawless town run by the notorious Soapy Smith and his band of thieves, the latter the starting point for the most famous gold rush trail of all: the Chilkoot. While there were numerous routes by which gold-seekers journeyed to the Klondike, it was the Chilkoot that was etched into the consciousness of the masses. Images of a never-ending stream of men labouring up 1,700 icy steps, burdened by loads of 50 pounds or more, through the snow-clogged cleft in the coastal mountains, have come to symbolize the challenges not just of the trail to the Klondike, but of life itself. Thousands made this ascent not once, but 20 or 30 times to haul the ton of supplies demanded by the North West Mounted Police who guarded the passes and patrolled the trails.

The stampeders faced challenges that would suck the life out of the most hardy: the lawless predators of Skagway; the numbing task of hauling 2,000 pounds of supplies, on foot, over a trail congested with ice, snow, and people; avalanche, drowning, and disease; exhaustion, failure, and heartbreak. Over the mountains and down the icy valleys came the "trail of 98," which reached the headwaters of the Yukon River. There, the hordes climbed aboard a hastily built, quirky armada of rafts, scows, and boats to float down 800 kilometres of windy lakes and rivers, through canyons and rapids, to reach Dawson City. Often, life-long friendships were torn asunder by the ordeals of the trail.

When they reached their destination, they found a bustling and rapidly growing city in a swampy hollow at the mouth of the Klondike River. There, scruffy Eldorado millionaire sourdoughs[1] rubbed shoulders with the newly arrived Cheechakos.[2] While some suffered from scurvy and deprivation, others displayed the symptoms of excess. These were heady times, during which the mundane was surrounded by the larger-than-life episodes of a grand adventure. And that is what it was – an adventure – for many of the new arrivals never made it to the gold fields or searched for gold. Just being part of the event was enough. And for those who sought their fortune in the Klondike gold fields, or stampeded to faraway creeks at the hint of a rumour of gold, most found frostbite and starvation, hardship and disappointment – for nothing.

Anyone who arrived in Dawson City found himself trapped there for the winter. By the time the stampeders arrived in the Klondike and searched for gold, it was too late to leave, for the summers are short in the north. Each man (there were few women in Dawson at first) had to build shelter for the winter, and then endure, for seven months, cold, darkness, disease, isolation, and monotony. For the lucky, nothing was beyond limits, and many of them lived extravagantly; for the majority, life was survival, the existence tedious.

For those who ventured forth into the hills to look for gold, there were many things to consider. Most of the rich ground was already claimed by miners in the first months following the discovery. Newcomers were obliged to work along the margins and beyond the perimeters of the ground that was already staked. Some were fortunate to secure bench claims (on the hillsides above the creeks) that the sourdoughs deemed worthless, and which often proved to be as rich as some of the creek claims. Everyone faced the same obstacles: transportation by foot, or dogsled if it was winter, to the creeks and tributaries where they had to thaw their way through many metres of frozen muck and gravel, and burrow, with pick and shovel, to the paystreak.[3]

1. A seasoned veteran prospector who has spent at least one year in the north.
2. Originating in the Chinook jargon, a cheechako is an inexperienced newcomer or tenderfoot.
3. A zone of gravel contained a dense concentration of unconsolidated gold.

E.H.ELWELL   ...ONANZA (BY FLASH LIGHT)

Miners working underground at Bonanza Creek
Photograph: Yukon Archives,
University of Washington photograph collection, # 1288

In the winter, this was achieved by sinking a shaft down to the bedrock upon which the dense gold had settled millennia before, and tunnelling (drifting)[4] along the paystreak to extract the concentrated gold-bearing ground found there. The frozen gravel and muck was softened by setting fires to thaw it. This material was hauled laboriously to the surface, where it accumulated in dumps[5] until the spring thaw, when the pay dirt was flushed through a sluice using the spring melt water to separate and capture the gold.

In the summer, working these shafts was too dangerous: melting permafrost[6] could collapse and crush a miner at any time, and lethal fumes from the fires used to thaw the muck accumulated underground. Instead, open pits were dug down to the bedrock, and the pay dirt hauled by hand to sluice boxes where the gold was captured in the riffles. In later years, the successful miners brought in steam-powered equipment that allowed more profitable enterprise.

Mining was always a gamble. Paystreaks meandered unpredictably through frozen gravel in the valley bottoms or in gravel deposits on hillsides. A man able to secure a claim would hope to find the paystreak on his ground, and would only know whether he was fortunate, or not, after months of back-breaking labour. If a claim proved to contain gold, the claim owner took on others to work for wages and mine the ground for him. In other cases, men formed a partnership, or lay,[7] and, for a percentage of the gold recovered, worked the ground on a claim belonging to someone else.

---

4.   A technique for extracting gold from permanently frozen gravels. First, a shaft was dug vertically into the ground until it reached solid rock (bedrock), Next, horizontal tunnels were sent out along the gravel/bedrock interface until the paystreak was encountered.

5.   A mound of gold-bearing gravel excavated from an underground paystreak during the winter, and stockpiled near the entrance to the mine until the gold could be extracted using the melt water from the spring thaw.

6   A zone of ground that remains perpetually frozen. It may start a few centimetres below the surface and extend down for a hundred metres or more.

7.   An arrangement in which a claim owner would agree to let a miner or group of miners work on his claim for an agreed upon share of the gold recovered.

Busy street in Dawson City

At the high point of the gold rush, more than 25,000 people went through the town.

Photograph: Yukon Archives, Vancouver Public Library collection, # 2095

The Klondike discovery occurred in 1896; the stampede occurred in 1897 and 1898. By 1899, the Klondike was old news. Having reached their goal, many hopefuls left the Klondike, returning home without having found their fortune. Others went off to a new gold strike in Nome, Alaska, on the coast of the Bering Sea. Gradually, the population of the Klondike dwindled from the 25,000 or more during the hey-day of the gold rush to a few hundred. A century later, however, gold mining is still the economic mainstay of the region.

What is the legacy of the Klondike gold rush? The cemeteries of the Yukon are filled with the bodies of those who had dreams and found misfortune. A few found wealth, and others returned home to live in comfort. Far more numerous were those who left with their pockets empty. If all the gold that was recovered during the heady years 1897 to 1899 was divided equally among all of those who participated in the gold rush, the amount would fall far short of the amount that they had invested, collectively, to reach the Klondike. The Klondike gold rush occurred during a period of economic depression, and sparked a massive infusion of money into the economy, spent by the thousands who thought that they might become rich for their effort. The continent emerged from the event on a healthy economic footing.

The gold rush brought about a rapid advance in the development of the Yukon Territory, which was officially formed by Parliament during the height of the gold rush, June 13, 1898, and left behind an infrastructure of supply, support, and governance that led to the continued development of the territory. Had it not been for the discovery of gold, development of this region would have been a slow and very gradual process.

The gold rush brought tremendous upheaval and disenfranchisement for the people indigenous to the region. The Han people of the Yukon valley were pushed aside and marginalized. Only a century later, as a result of land-claim settlement, are the Tr'ondëk Hwech'in emerging as important partners in the running of the region.

Perhaps the most lasting legacy of the Klondike gold rush is the impression that it left in the public mind. Tens of thousands from all walks of life and from all parts of the globe embarked on the adventure that was the gold rush. They shared an experience, which they all faced, rich or poor, on

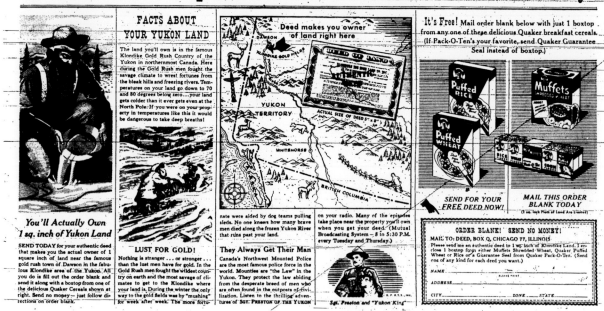

a relatively similar footing and which left its mark indelibly etched in their memories. Subsequent to their adventure, they lived their lives, Pierre Berton explained, as though they were "climbing a perpetual Chilkoot." Hundreds of books have been written about the gold rush that recount the individual experiences of those who were there. Thousands of photos were taken that capture some of the most poignant events of the odyssey – the stream of men filing up the "golden stairs," the armada of gold seekers floating down the Yukon river, the crowded dusty streets of Dawson City in its heyday, and the miners mucking the gold out of the Klondike streams. These images have become icons of the current generations, symbolic of the epic quest of a century ago.

The adventures of the gold rush were also captured in popular literature in the writings of such people as Rex Beach, Jack London, and Robert Service. Their works have ensured that the Klondike gold will not be soon forgotten. Words like Klondike and Chilkoot evoke images of gold, adventure, challenge, and the north. They are found everywhere in our vocabulary. There is a Klondike

The Quaker Oats Company had one of the most profitable campaigns in the history of North American advertising. In the 1950s, it sold boxes of cereal containing a one-inch-square property title to a mining concession in the Klondike.

Photograph: copy of an original from the *Seattle Times*

ice cream bar, and a Chilkoot automobile, a town, streets, and schools named Klondike. There is a card game called Klondike; the City of Edmonton celebrated "Klondike Days" for years, and one of the most successful product promotions in North American advertising history came about when Quaker Oats packages were sold containing a deed to a square inch of a gold claim in the Klondike.[8]

The Klondike gold rush will long be remembered for the images from the trail, the hardships endured, the stories told, the dreams dreamt, and the adventures lived.

Bibliography

Berton, Pierrre. *Klondike: The Last Gold Rush, 1896-1899.* Toronto: Anchor Canada, 2001.

---

8.   This successful marketing enterprise ended in legal procedures invalidating owners' property rights for non-payment of taxes. The Canadian government seized the land in 1965 for an outstanding $32.70 tax bill. [http://www.yukoninfo.com/klondikebiginch.htm]

Gold Rushes in Quebec

JOSÉ LOPEZ ARELLANO

Anthropologist, head of the exhibition's Research Committee, Musée de la civilisation, Quebec City

Translated by Joan Irving

FRANÇOIS I, KING OF France, realizing that Spain's Golden Age was being propelled by the influx of precious metals from the Americas, commissioned Jacques Cartier "to discover certain islands and lands where it is said that a great quantity of gold, and other precious things, are to be found."[1] The gold and diamonds that Cartier carried back from America turned out to be fake, thus forestalling a rush to the new French colony in America.

Quebec didn't have a real gold rush until 1863, following an accidental discovery of the metal in the Beauce region. According to the oral tradition, Clothilde Gilbert found a gold nugget the size of an egg in a riverbed one day while herding her cows back to the barn.[2] The Plantes, du Loup, and the Cumberland rivers were soon being explored for gold.

Many of the prospectors who flocked to the area had taken part in the California Gold Rush. They, with their wealth of experience, were convinced that the topographical evidence pointed to the existence of rich gold-bearing veins in the sands of the Beauce's rivers. Great was their disappointment!

Although the region's gold production increased substantially after 1867, the major gold veins were too deep and largely inaccessible to these prospectors, who lacked the necessary specialized mining techniques. Gold mining in the region gradually decreased.[3]

In the 1920s, farmers as well as geologists working for the railways, guided by local Aboriginals, discovered gold deposits in the sub-soil of the Abitibi region. When this discovery was announced, hundreds of prospectors from all over Quebec, Ontario, and the United States were struck by gold fever and boarded trains for the area. Pausing only long enough to erect their tents and

Group of Quebec gold prospectors, c. 1896
When they left, their hearts were full of hope and pride.

Musée de la civilisation, gift of Georgette Doré and Yvette Doré
Photograph: Idra Labrie, Perspective,
Musée de la civilisation. No. Ph2007-0007

log cabins, they set out in search of any promising vein. Thus, thanks to the railways, prospecting and mining became the engines of colonization in northwest Quebec.[4] Deep in the forest, several towns went up: Val-d'Or, Bourlamaque, Cadillac, Malartic, Pascalis.

The mines attracted European immigrants: Irish, Germans, Poles, Ukrainians, and Italians. Like those during the California and Klondike gold rushes, these immigrant miners were intent on making a small fortune, then sailing for home and buying themselves a small parcel of land. Only a small minority of them settled in Quebec's northern towns.

---

1. Cartier, in Biggar, 1993.
2. *Cf.* Paquette, 2000; Pomerleau, 1996; Vallières, 1989.
3. *Cf.* Paquette, 2000: 25.
4. The new northern frontier was rich in forestry resources and was also the gateway to James Bay.

Pocket watch belonging to Anaclet Dupuis, a Quebecer who went to the Yukon in 1897–98 to seek his fortune
Private collection, Photograph: Annie Beauregard

Then, with the decline in the price of gold of the early 1970s, the least profitable veins were abandoned and mines were closed, threatening the very existence of the northern towns. Despite the expansion of forestry and the development of industrial parks, hundreds of miners had no choice but to leave the region.[5]

During thirty glorious years (1945–75), thousands of Quebecers travelled the Highway North, drawn by the prosperity of towns founded by mining companies: It was cold, it was far from everything, and it was the next thing to paradise.

In about the mid-1980s, mine production rose once again, in conjunction with the development of new industrial applications for gold. Mines were also becoming more efficient, and less costly to operate, because of improvements in mining techniques. Since that time, there has been constant growth in four sectors of gold exploration: prospecting, geophysics, geochemistry, and geology. Quebec's gold-mining industry has gained an enviable reputation around the world in the fields of diamond drilling, methods of geochemical sampling, knowledge of mineralization processes, and the creation of metallogenic models.[6]

Bibliography

Armstrong, R. 1984. *Structure and Change. An Economic History of Quebec*. Toronto: Gage Publishing Limited.

Cartier, Jacques. 1924/1993. *The Voyages of Jacques Cartier*. Translated by Henry P. Biggar. With an introduction by Ramsay Cook. Toronto: University of Toronto Press.

Daoust, A., G. Gaudreau, and K. Reilly. 1998. *La mobilité des ouvriers-mineurs du nord ontarien et québécois* 1900-1939. Série monographique en sciences humaines, xii-141. Sudbury, ON: Université laurentienne/Institut franco-ontarien.

Gauthier, J., and P. O'Dowd. 1991. "Réflexions sur quelques innovations qui ont marqué l'exploration minière." In *Les innovations dans le monde minier québécois*, ed. Gilles Saint-Pierre, Gratien C. Gélinas, and Marcel Vallée, 15–27. Montreal: Gaëtan Morin éditeur.

Hamelin, J. 1976. *Histoire du Québec*. Montreal: Privat-Édisem.

Paquette, P. 2000. *Les mines du Québec 1867-1975. Une évaluation critique d'un mode historique d'industrialisation nationale*. Outremont QC: Carte blanche.

Pomerleau, J. 1996. *Les chercheurs d'or. Des Canadiens français épris de richesse et d'aventure*. Quebec City: Éditions Jean-Claude Dupont.

Porsild, C. 1998. *Gamblers and Dreamers: Women, Men and Community in the Klondike:* Vancouver: University of British Columbia Press.

Saint-Pierre, G., G. Gélinas, and M. Vallée. 1991. *Les innovations dans le monde minier québécois*. Montreal: Gaëtan Morin éditeur.

Vallières, M. 1989. *Des mines et des hommes. Histoire de l'industrie minérale québécoise: des origines au début des années 198*. Quebec City: Les Publications du Québec.

---

5. The increase in northern mining, the founding of new towns, and the great mobility of people parallelled the increase in industrialization in Quebec and its growing integration with the economy of the United States. *Cf.* Paquette, 2000: 152.
6. Gauthier and O'Dowd, 1991.

Zidor le prospecteur

Du plomb dans la tête
De l'argent plein les poches
Du fer dans les bottes
Et de l'or sur les dents
Le v'là qui s'arrête
Le v'là qui s'approche
« Si vous faites un pot'
J'mets cent piastres dedans ».
[...]
De l'argent sur les tempes
Du plomb plein les bottes
L'or dans la caboche
Et rien sous la dent
Le v'là qui r'décampe
Qui remet sa calotte
Va chercher des roches
Pour leur voir le dedans

Gilles Vigneault

Roméo Céré, Quebec gold prospector (Desmaraisville)
Photograph: Jean-Claude Labrecque, taken from the documentary *Une rivière imaginaire* (written and directed by Anne Ardouin), ©1993

Fool's Gold

YVES CHRÉTIEN, HÉLÈNE CÔTÉ, RICHARD FISET AND GILLES SAMSON
Archaeologists, Commission de la capitale nationale du Québec

Translated by Joan Irving

Jacques Cartier Discovering the St. Lawrence River, 1535

J. A. Théodore Gudin, 1847
Chateau Versailles and Chateau Trianon, Versailles, France
Photograph: Réunion des Musées Nationaux/Art Resource, New York

IN OCTOBER 2005, DURING excavations initiated by the Commission de la capitale nationale du Québec, the site of Jacques Cartier and Jean-François de La Rocque de Roberval's settlement at

Cap-Rouge was located. This discovery provided irrefutable evidence of the aims of these sixteenth-century explorers: to locate the riches of the land, in particular, that most precious of metals – gold.

In August 1541, upon disembarking at Cap-Rouge, the famous navigator from Saint-Malo marvelled at the resources of the substratum, which was rich with diamonds and gold.

> . . . Adjoyning [sayd Fort] we found good store of stones, which we esteemed to be Diamants. On the other side of the said mountaine and at the foote thereof, which is towards the great River is all along a goodly Myne of the best yron in the world, and it reacheth even hard unto our Fort and the sand on which we tread on is perfect refined Myne, ready to be put into the fornace. And on the waters side we found certaine leaves of fine gold as thicke as a mans nayle.[1]

Cartier returned to France with barrels full of these minerals, intending them as gifts for François I. Analyses undertaken in France concluded, however, that he had been mistaken as to the nature of the "riches." The diamonds were just quartz and the gold was iron pyrite, or "fool's gold." Pyrite is shiny and yellow like gold, and is easily mistaken for it.

"Diagnostic" artifacts found at the Cartier-Roberval archaeological site clearly show that the French "prospectors" had undertaken more than a simple visual identification to ensure that they had found precious metals. Refractory clay containers that had been used as crucibles were dug up and identified, indicating that the French sailors who visited Cap-Rouge in the sixteenth century were capable of chemical analyses.

Crucibles used for metallurgical testing have to withstand very high temperatures because the substance being tested is heated with other metals in order to separate the ore from its matrix. This is why refractory clay is used in their manufacture.

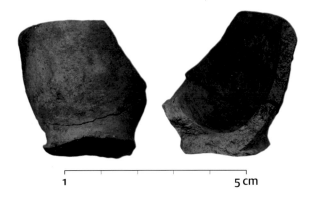

Shards of a small crucible

Photograph: Yves Chrétien

The crucibles found at this site are all small in size, about 3 centimetres high and about 4 centimetres across. They are fluted with a pedestal base, to ensure their stability.

The method of chemical analysis from the period consisted of grinding the mineral to a powder, pouring it into the crucible, and then adding powdered lead oxide, to draw the metallic component of the mineral to the bottom of the container. The crucible and its contents were then heated in a very hot oven to bring the mixture to its melting point. This operation separated the mixture into two parts: the slag that rose to the surface and the metallic mixture and lead that settled at the bottom of the container. After allowing this to harden, the crucible was broken and the contents removed. A crucible was used only once, which accounts for the many such shards found at the test site. Other measures, requiring the use of other types of containers, were then undertaken to separate the lead from the precious metal.

The fifteen fragments from six different crucibles uncovered at the Cartier-Roberval site corroborate the statements of the two explorers that one of the purposes of their voyage was to assess the mineral resources of Canada and to undertake metallurgical experiments right on the spot.

How did they do this, and what did they really find during their tests to verify the gold content of the ore? This is what the archaeological work at Cap-Rouge is trying to determine. The experts are hoping that the fragments still bear minute traces of the chemical elements used in treating the ore, and that it will be possible to identify them by analyzing the fired clay.

1. *The Third Voyage of Discovery Made by Captaine Jaques Cartier, 1541,* p 98. Wisconsin Historical Society Digital Library and Archives, 2003. www.americanjourneys.org.

Clandestino
by Manu Chao, from the album *Clandestino*, 1998.

Pa una ciudad del norte
Yo me fui a trabajar
Mi vida la deja
Entre ceuta y Gilbraltar
Soy una raya en el mar
Fantasma en la ciudad mi vida va prohibida
Dice la autoridad
mano negra clandestina
peruano clandestino
africano clamdestino
marijuana ilegal
Solo voy con mi pena
Sola va mi condena
Corre es mi destino
Por no llevar papel
Perdido en el corazon
De la grande babylon
me dicen en clandestino
yo soy el quiebra ley
[...]

CHAPTER **6**

GOLD SEEKERS OF TODAY

Gold Mining on the Pacific Coast of Colombia: Río Timbiquí

Michael Taussig, Ph.D. Anthropology, Columbia University, New York

Gold Mining on the Pacific Coast of Colombia: Río Timbiquí

MICHAEL TAUSSIG

Anthropologist, Professor, Department of Anthropology
Columbia University, New York

Excerpt from his book *My Cocaine Museum*,
chapter 9, "Mines," The University of Chicago Press, 2004

IN OUR TIME: ENRIQUE, aged fourteen, takes me to the mine where he works after school with his sister, Wendy Sulay, aged eight, and his brother Walter, aged ten. They are part of a group of some twenty children and adults, largely women, who divide the profits unequally, if there are any to be divided. The owner of the land and the owner of the water pump take 50 percent between them. The kids get a small amount of cash at the end of three weeks. If there is no gold, nobody gets anything. I am told it is not uncommon to wait years ...

We walk past the outlying houses. Already it feels like we have passed an important division between town and country. Enrique bounces a soccer ball. We are joined by Javier, who teaches English at the school and wants to practise speaking it with me. Dairo, aged twelve, who refused point-blank a year ago to go back to school, is with us too. We follow the stream upward, passing through the shade of breadfruit trees with their sharply etched leaves, splashing through the clear water, feet searching for position on the round stones, Enrique leading, lithe and sure-footed. The water dies out but the stones continue in the dry creek bed. We walk through a grove of plantains. An old woman with no clothes above her waist is crouched on the ground. Old felt hat and

a necklace of beads, a bunch of plantains on the ground to carry to the village. Now it is slippery underfoot. A tree trunk obstructs the path. We pass a hole in the ground with old beams poking through rotting vegetation. It is an old mine.

A little farther on we come to the mine where Enrique works. First we see the wooden sluice standing in an orange-sandy-bottomed stream running by the side of coconut palms and breadfruit trees. The mine is what is called a *pozo*, or hole, that goes straight down for about twenty feet. From the bottom, small tunnels are excavated horizontally in a radial pattern. The hole is elaborately buttressed by criss-crossing branches and tree trunks. The French method was to cut horizontal tunnels in from the river banks and rely on gravity drainage, but now, with pumps, you can dig straight down anywhere and then radiate out in a star pattern underground, although these mines extend nowhere near as long underground as the tunnel mines. For example, Leopoldo's tunnel mine, an extension of a French one, is some 1,300 metres long and at times 80 metres below the surface.

Enrique jumps down into the hole. Dairo follows. Trapeze artists making their way into the bowels of the earth. You can barely see the bottom. It's full of water down there. About one and half metres of water, says Enrique. Nobody is down there today. They are waiting for others to cut more beams. Enrique worked a full year in that mine and never once saw gold. His father was working in another mine for five and half months and at the end got a dollar and fifty cents.

◄ Colombian miners in a artisanal gold mine, Quibdo, Colombia

Photograph: Jeremy Horner/Corbis

He's never going into a mine again, he says. But Enrique, Wendy, and Walter keep at it.

Dairo's father threw him out of their home a year ago for being disobedient, and he now lives with an uncle who teaches school and drinks too much *biché*. Every day he drinks *biché*. The uncle is small and beautiful, delicately featured with morose red-lined eyes, and he steers an uneven course up and down the street, swaying with the play of the cobblestones. Distilled under the stilt-houses along the adjoining Río Saija, *biché* is scary stuff; potent as dynamite, cheap as air, it is, as its name suggests, rawness on its way to something else. In the bottle it smells like lighter fluid. Exhaled it gives off an aroma like honey mixed with vomit that hangs onto the person drinking it like a halo. Too much and you get to be a *degenerado*, like Dairo's uncle. He walks unannounced into people's houses, breaks into their conversations with absurd questions, begs for rice and sugar, and refuses to leave. Although rich by local standards with a government cheque equivalent to 350 dollars a month on account of the many years he's been teaching, he's always broke and Dairo is always hungry, the only kid who asks me for money. His uncle's wife drinks a lot also and decided to live elsewhere, so young Dairo and his desperate uncle are left to cook for themselves.

People tell me his uncle is Gustavo's not-so-secret lover. Yet while Gustavo is noisy, demonstrative, and somewhat feminine in his gait and garb, pursued by raucous young boys teasing him to death on his way to pan gold with the women,

Dairo's uncle is morose and withdrawn, a red-eyed phantom trailing honey and vomit. No wonder people say Dairo is an odd kid. And no wonder he is the only kid who doesn't go to school, where his uncle presides. A year later I heard he'd split and was working in a refrigerator plant way down in the mouth of the river, gutting and cleaning fish. It must be lonely down there in a decrepit building about to collapse into the gloomy estuary, no other houses around, just mud, soporific heat, mosquitoes, and the occasional surprise visit by the guerrilla hungry for fish. At least Dairo will probably get something to eat there.

Sometimes the village seems like the place to where the Pied Piper of Hameln spirited all those kids with his magic flute. Nothing but kids shrieking with laughter, swarming like sardines flowing in and out of the street. What joy. What light. What bubbling life. They are either going to school or coming back from school, neat and polished in their school uniforms and shiny black shoes. In the high school, boys and girls wear the same colours, a crisp white shirt and grey pants, for boys, a skirt for girls made of a light grey checked material with a thin red stripe. There is a lot of argument going on about changing this to blue jeans. A uniform is a huge investment. In 1998 there were eighty-four pupils in the high school and close to four hundred in the grade school. School sets the rhythm of the day, no less than the hopes for a future in the wider world.

When the children are at school, the village is somnolent, basking in the heat and ready to expire. All you can hear are the hot cobblestones grumbling and getting ready for the next bout of noisy rattling when the kids romp home from school. The teachers seem to find school too hot. Much of the time you see them chatting under a tree in the little park opposite the school, the park with the statue of Justinano Ocoró, who left school at age twelve and, as they say, abolished slavery. The rector of the high school gets the equivalent of US$750 a month (plus a mighty pension upon retirement); a teacher on a yearly contract in the grade school, US$100 – staggeringly high incomes compared with the rest of the river. Schoolteachers wear T-shirts with the name of their school and river emblazoned across them as names of football teams. In the canoe going upstream, a man carefully stows away a wooden plaque made by a prisoner in Santa Barbara. The prisoner made it for his teacher in Santa María and it has a dedication engraved to her. To be a teacher here is to be a god.

Despite the effervescence that is the children bubbling with life, this is a ghost town in that so many of its people don't live here anymore and the ones who do are marking time, like so much of Colombia. When in mad discord the church bells ring the Day of the Dead, they ring as much for this emptiness as for the ghosts of remembrance. Take Yanet, aged thirteen, who lives with her mother, her mother's parents, and her four half-brothers and sisters in a two-room single-story wooden house at the end of the village. It is dark as we talk and I scribble notes by candlelight, tracing out a spidery genealogy.

They have no money and live off plantain and boiled *yuyo* gathered from the forest. Yanet works for Lilia before and after school until eight o'clock at night, seven days a week. Instead of money she gets meals and her school uniform. Lilia is able to afford this because she runs a government-funded crèche, which is likely to be phased out as the present government cuts welfare in accord with IMF and World Bank promotion of market forces. Yanet's father lives in the port of Buenaventura, way up the coast. Her stepfather is separated from her mother and lives with another woman elsewhere in Santa María. Her mother's father,

Apolinar, has four siblings in Santa María and five in Cali. Four out of five of her mother's mother's siblings live in Cali, the other in Buenaventura. The women in Cali work as live-in servants, the men in construction, if there is any, which there isn't. This is a hollowed-out village. The census I did in 1976, some fifteen years before talking with Yanet, showed that for every people born here, only four were still resident. The situation now must be even more extreme.

The village vomits out people. It has no future. It was once alive with Indians making golden fishhooks. Then it was an outpost of empire, an African slave camp sending gold to Panama to await the Spanish fleet. Now it is a scar on time. Most of the coast has this eerie artifactual feel. Nobody would live here unless they were born here, like the Indians, or forced to, like the slaves, or shipwrecked, like the whites on the beach at Mulatos. It's high time everyone cleared out, which, in truth, seems to be everyone's dream.

Let me tell you about *yuyo*, what Yanet's family survives upon together with boiled plantain. As far as I know, *yuyo*, large-leafed and thorny, is special to the diet of the Río Timbiquí. Nowhere else on the coast, nor for that matter in rural Colombia, is anything like it eaten. And no peasant I know of in the interior eats greens. Moreover the tiny thorns running along the surface of the leaves produce welts on the skin. Yet in Santa María *yuyo* is food. From conversation with Daniela Zuñiga, who as a young girl helped her mother cook for the New Timbiquí Gold Mines in Santa María in the 1920s, I got the idea that the French engineers used *yuyo* as a substitute for lettuce. It seems like the river was humming those days with flotillas of canoes carrying wines and cheese from France. No doubt I exaggerate. But what would a French lettuce look like after a month or more at sea and the rigors of the ascent up the Timbiquí?

This was a colonial fantasy I hung on to. It gave me pleasure to think of those robust French engineers so far from home, so willing to risk their health and possibly their lives for gold, but unable to do so without the customary pleasures of the table. It made them human to me. And it made them real. It brought their ghosts into the present, along with their stiff-necked presence in the occasional family photographs you glimpsed in homes along the river.

When I actually get to see *yuyo* and taste it, I realize I am way wrong. It is more like cabbage than lettuce and has to be boiled to softness – otherwise, it's too prickly. Here I am like an eighteenth-century botanist, or a privateer like Woodes Rogers in 1709, trying to understand something new in a new world by forcing it into my categories, lettuce or cabbage, as when Woodes Rogers saw his first sloth on the island of Gorgona and said it was a monkey, surprised it took all day to climb the mast that a monkey would scamper up in a minute. *Yuyo* has recently become a health-food speciality in Cali. Truly, ours is the time for the great awakening, the *apertura* of the coast, as the young interpreter for the Russians says.

In our time: Enrique, now aged seventeen, takes me to El Limbo. It fully deserves its name. Up hill and down dale we slither and slide. I find it almost impassable. I lose a shoe in a swamp. The forest presses in dark and dank with its brilliant shades of green. Sweat pours. At times we have to haul ourselves up with vines, finding toeholds in notches cut into slippery *chonta* palm trunks. Finally we break through the forest onto an immense beach of white and grey stones by the edge of a tawny river running fast and smooth in a graceful curve. There is an old man, don Julio, bent over by the weight of a bunch of plantains he is carrying toward a tiny shelter of branches built on the edge of the forest.

Soaring opposite on the other side of the river is a cliff dripping water and exotic ferns with leaves shaped like tears streaming down its face. Twelve feet above the rushing river is the dark hole of the mine. A young man picks us up in a tiny canoe with an inch of freeboard. It wobbles with the slightest movement. If you grip the sides too hard, it leans into the river. Be still, don't show your fear, says the river. We drift with the current, get out, then have to climb the cliff on slippery wet indentations and inch along sideways till we get to the mine opening, where there is a slight protuberance that allows me to swing onto the floor at the entrance to the mine. Agida is there, smiling. "Watch out for the buggy!" she yells, and presses me against the dripping wall as a young man comes teetering out of the darkness with a

Man poling his pirogue against the current
Photograph: Michael Taussig

wheelbarrow full of slimy grey rocks and mud. There is a pungent smell due, they say, to the freshly cut timber supports.

The principle of the human chain: Down below us braced against the current of the river are thirteen people passing buckets of stones and mud from hand to hand with lightning speed down the cliff to the wooden *canelón*, or sluice, standing in water. Some people stand in the river; others cling to the cliff. What care they take! Even rubbing dirt of the rocks brought out of the mine so as to get every possible piece of dirt washed in the sluice below. How the women on either side of the sluice churn the gravel pushed forward by the force of the river, up to the elbows at times! How they move those *bateas* in the river, swirling the fine gravel for the final spin! The massive strength and endurance of those bodies, picking out the heavy stones, hurling them into the river. Agida at it all day long.

Of the seventeen people, ten are kids, ages ranging from six to thirteen. Three men chip away at the rock face three hundred feet within the mine, searching for what they call the *viejo canal*, the old riverbed, where the river ran in ancient

times before the Flood. Except for Agida, they are all kin-related. No gold yet, but I see hopeful glints in the *canelón* presided over by the owner's mother and wife. His name is Carlos Arturo Benté and he invites me inside the mine. "It's your home," he beams. "It's safer than an airplane!"

That night I go to Agida's house. It is full of children spilling out into the street watching TV. I find her at the end of the corridor in the kitchen preparing dinner after her day at the mine. It is pitch-black in there as she can only afford one electrical outlet, either the TV or a lightbulb, costing the equivalent of US$3 per month for the electricity. It is pitch-black in that kitchen, like the mine.

Omar's mine, which is really his mother's, used to be a French mine. It's easy to find, ten minutes' walk from the village by a stream shrouded by breadfruit trees from which the ear-splitting noise of the pump emerges, along with endless lengths of black-ribbed rubbery hose. As you enter the darkness with a carbide lamp brighter than any electric light, innumerable chip marks made by the pick enclose you. I stop, scared to go farther.

Years later I did enter a mine, its entrance a weeping hole in a cliff that formed the riverbank, the floor being no more than a foot above the level of the river rushing by. A small stream fed into the river with a little sandbank on which a group of muddy kids and women were waiting in the rain for earth to be carried out, it being their job to wash this for gold. The miners were too poor to have anything but candles and because the tunnel took a right-angled turn it was pitch-black most of the way. It became clammy and hot and progressively harder to breathe. I couldn't imagine what it would be like to heave a pick and iron bar all day at the pit face. My flashlight revealed intricate roof supports as we sloshed through the sludge. I felt I was choking and then saw the light after turning the bend. Two elderly skinny guys stood there scratching their heads, surrounded by boulders about two feet in diameter. An unholy mess. Somehow I had expected a nice flat floor and a neat right-angled wall of stone at which they would be chipping away. A real tunnel, you might say. Instead there was this gruesome disembowelment of mother earth with everything at sixes and sevens, oozing muddy-water and nameless fluids.

As I turned to go back, I began to feel curiously at home and cosy in the mine, perhaps because I knew I was on the way out and could start to reflect on this as an experience that I now hand over to you. This is the basis of many theories of history, personal no less than of the worldly. At first the human being is so immersed in reality, in this case horrific, that she or he has neither consciousness nor self-consciousness. There is no Other, just the interior of the pitch-black mine penetrating your being. Then comes the second part of the story. Evolving differentiation enters the scene. Subject

Folk miner in front of a mine entrance
Photograph: Michael Taussig

peels off from object allowing for consciousness of self. Aha! I am having an experience!

Plato said this occurred in a cave, too. But there was light in his cave, casting shadows that led people astray. They were having misleading experiences, he said, and it was his job to free them from illusion and lead them out of the cave. The great Hegel rewrote this story of experience with his parable of master and slave. While it is the master who make decisions, like Julio Arboleda, the slave works at the rock face with bare hands or by the river bank with a wooden bowl and enters into an understanding of the nature of nature that subtly intertwines being inside as well as being outside. The slave's selfhood emanates from this interaction with the physical world, framed by servitude to the master. But the master's consciousness sets him apart from nature, and he has experiences like looking out a train window or looking at gold.

In reference to this system of tunnel mining introduced by the French, and which subsequently spread to adjoining rivers like the upper Saija, the Guapí, and the Micay, Robert West wrote in 1960 that in inexperienced hands this is a most dangerous technique, and that frequent collapse of the tunnels has killed many due to faulty timbering.[1] I hear that a villager died this way a year ago and that this is frequent. William Amú tells me that about twenty people have died from mining accidents in his memory over some forty years; Lilia says thirty, this out of a total population of twenty-five hundred, a high mortality rate, indeed, far higher than the homicide rates in Cali, for instance, considered among the highest in the world.

I am outside one of the new style of mines cut straight down. The earth is shaking with the roar of the pump and water. Blue plastic sheeting lies everywhere. There are tiny shelters made of banana leaves, women underneath lying down half asleep or legs straight out in front, lit end of cigarette inside the mouth, waiting for the men to start excavating. Then they form a human chain, passing the earth dug from the mine in flexible rubber buckets, throwing each bucket from hand to hand until it gets to the two women at the sluice into which gushes the water the pump has extracted from the mine. Their hands move like lightning sorting big rocks from small, gravel from fine gravel, heaving buckets of stones out of the sluice, then stemming the flow of water so as to sift more thoroughly. Later in the afternoon they will wash the fine dirt using a *batea* to get the even finer, black pay dirt, called *jagua*. In many areas of South America today, just as in colonial times, mercury is used to separate gold from the pay dirt, with terrifying consequences for human health and the environment. But I never heard of its use here. Robert West reported in the early 1950s that most folk miners on the coast then used a trick acquired from the Indians, using a glutinous sap from trees like the balsa tree to create a surface tension to which the iron in the pay dirt clung, while the gold drifted to the bottom of the *batea*, but I never saw this.[2]

In fact, I never saw gold itself but on a couple of occasions, swimming like tiny stars in water at the bottom of a split coconut shell, which is how women keep it during the day out in the forest by the streams they are working. Why do they keep it like that, like little fish swimming in water? Why do we always find this intimacy of gold and water? Another time I saw a strong young woman, her five-year-old child watching from the bank, dive repeatedly with a stone tied to her back. In an eddy by her side bobbing up and down was a split coconut shell floating precariously on a torn-off piece of plantain husk. I keep thinking of the nonchalance and fragility embodied in this tiny craft destined to hold whatever gold she might find.

It sums up all I feel about Santa Mariá.

Tunnel mines are very rare in the Pacific Coast, other than on the Río Timbiquí where they were introduced by a French and British mining company early in the twentieth century which effectively sealed off the area and ran its own "republic" with its own jail, police, and currency, so as to ensure a sufficient number of workers. The company left the area in the early 1930s.

Today, most mining in terms of numbers of people would be that done by women and children washing gold in the streams, sometimes diving as well with a heavy stone strapped to their back. That would be the more typical mode of gold mining in Colombia today, as it has been for four hundred years. Nearly all gold miners in Colombia are "folk" miners, descendants of the African slaves who worked for more than three centuries for white men until the abolition of slavery was completed in 1851. To the great distress of their previous owners, the ex-slaves, by and large,

---

1. Robert West, *The Pacific Lowlands of Colombia: A Negroid Area of the American Tropics* (Baton Rouge, Louisiana State University Press, 1960.) p. 178.
2. Ibid., p. 178.

Child looking for gold

Photograph: Michael Taussig

refused to continue working for a master even if good wages were offered. They wanted their own land and they settled into a subsistence economy far from towns, usually along rivers accessible only by dug-out canoes. They grew their own food: plantains, manioc, and corn, hunted animals in the rain forest, and fished in the estuaries. Today this pattern still holds as many of the gold miners live far from markets. Upriver where the miners live there is little game and practically no fish anymore. The food is monotonous and insufficient. Most people yearn to leave for the cities of the interior.

Some 10 percent of Colombia's 50 million people are classified as of African descent. They live in the hottest and least salubrious parts of the country.

Most folk miners live today on the isolated Pacific coast along fast running rivers that rush down the western mountain chain that lies close to the sea. There are no roads there. The rainfall is incredible, the highest or second highest in the world. Mangrove swamps run four hundred miles. This is no "tropical paradise." It is extremely hot and humid. Malaria is common. The people have little to no money. In the gold mining villages I know of, a dollar a day is an average income. To the visitor, it feels like the end of the earth.

This is what makes the area now a haven for the cocaine trade as growers of coca move away from the areas killed off by USA-enforced spraying of the upper Amazon basin. Unwilling or unable to curb demand within its borders, the US seeks to curb production in Colombia. After more than a decade and more than a billion dollars from the US treasury, there has been no decrease in the amount of cocaine produced while the amount of violence created by legislation outlawing the drug has been absolutely catastrophic to Colombians.

Because there is so little gold left that can be reached by folk miners, some miners or their adolescent children now fearfully participate in the coca business penetrating the coastal rivers as mutually hostile groups battle for control of this lucrative crop. The miners face the possibility of massacres by paramilitaries, often allied with the national army. There are also onerous demands by the *guerrilla,* now more of a business organization than a left-wing movement. And it is likely that the government will subject the miners' fields and the surrounding rainforest to chemical spraying too.

Colombia was far and away the chief gold producer of the Americas and the vast Spanish empire. The fabulous wealth that the colony provided Spain rested on the backs of slaves. None of this is even officially noted. It is as if history has been erased. Why is that? And why is it that the ghastly toil is likewise effaced in the glitter of gold? Truly, as the miners say, gold belongs to the devil.

The only public bench in Sioux Junction . . . faces away from the river, the Wicked Sarah, where thirteen children had drowned since the town's founding at the beginning of the century. The last drowning – that of the little L'Heureux boy, who was searching for gold in the river like the prospectors who had made the city rich – went back to the days when the town was threatened with closure. On the day before he drowned, the boy had learned that the town was about to be closed because the nearby mine and forest had nothing more to give. The child got it into his head that if he found gold in the Wicked Sarah the prospectors would come back, as would the bosses, the Americans, the workers, the many families who'd left, and the penny-candy seller who'd moved away one month earlier. At school, he'd learned that the founders of the town had made fortunes after discovering gold nuggets in the sand of the Wicked Sarah.

Poliquin, Daniel. 1987. L'Obomsawin.
Montreal: Bibliothèque québécoise, p. 9.

CHAPTER **7**

GEOLOGY

Gold: From Source to Final Products

JAYANTA GUHA, Ph.D. Geology, Professor emeritus in Earth Sciences at Université du Québec à Chicoutimi

BENOÎT DUBÉ, Ph.D. Geology and Engineering, Research scientist, Geological Survey of Canada,
Natural Resources Canada

The *Buisson d'or*

MICHEL GUIRAUD, Professor of mineralogy and Head of Collections, Muséum national d'histoire naturelle, Paris

Mineral Exploration in the Twenty-first Century

JEAN-MARC LULIN, Ph.D., geologist, President and Chief Executive Officer, Azimut Exploration Inc.

Agnico-Eagle Mines Limited

EBERHARD SCHERKUS, P.Eng., President and Chief Operating Officer

The Éléonore Project: Creating the Future!

CLAUDE-FRANÇOIS LEMASSON, P. Eng., General Manager – Projects, Canada & USA

Gold: From Source to Final Products

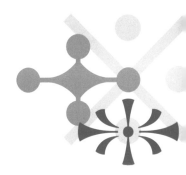

Jayanta Guha

Geologist, professor emeritus in earth sciences at Université du Québec à Chicoutimi

Benoît Dubé

Geologist and engineer, research scientist, Geological Survey of Canada, Natural Resources Canada

Translated by Joan Irving

Gold, the most sought-after and loved of metals in the history of humanity, is the result of a variety of geological phenomena that take place both deep within and on the surface of Earth. The discovery of the famous gold mask of the Egyptian pharaoh Tutankhamun in his tomb illustrates the great fascination of this precious metal through the ages. The following is a summary of gold's physiochemical characteristics; geological and geographic location and distribution; exploitation, extraction, and transformation methods; uses and applications; and repercussions on our lives.

Physiochemical Properties and the Natural Distribution of Gold

Gold (chemical symbol Au, atomic number 79) is a precious metal that is a shiny yellow colour. It is found in the group IB of the periodic table of metals. Its colour varies according to the percentage of silver commonly associated with it: the presence of silver lightens the gold colour. Gold's specific weight, 19.32 grams per cubic centimetre, is an indication of its great density. It's greatest property is its ductility: gold is malleable at room temperature, although this diminishes in the presence of other metal alloys. Very resistant to corrosion because it does not combine with oxygen, gold is attacked only by aqua regia, that is, nitric acid mixed with hydrochloric acid to form gold

Industrial waste, Andes Cordillera, Chile

Photograph: Marcelo Riveros, 2007

Very high content gold ore from the Red Lake mine in Ontario

Photograph: Benoît Dubé

chloride. Gold also reacts with potassium cyanide to produce potassium dicyanoaurate which, when dissolved in water, becomes very toxic. Earth's crust is made up of 0.005 parts per million (ppm) of gold, compared to 50 ppm for copper and 0.07 ppm for silver (Boyle, 1987).

In nature, gold is found mostly in a native state and as the main component in alloys with silver, copper, and metals like platinum, etcetera (Boyle, 1987). Like silver, gold acts with tellurium to form tellurides such as the following minerals: sylvanite, calaverite, petzite, etcetera. Gold also forms the minerals antimony, selenium, and bismuth.

Gold ore is found in a variety of forms. The most common and sought-after have long been gold nuggets and flakes. These occur mainly in the sediments of waterways and in sand or gravel in rivers and streams, as placer deposits (sand and gravel) and paleoplacer deposits (rocks). The biggest gold nugget ever found in Quebec, the McDonald Nugget, was taken from the Gilbert River in the Beauce region in 1866. It weighed 45 ounces. Paleoplacer deposits in South Africa

(Witwatersrand), discovered in 1884, remain the largest gold deposit ever found, with over 78,000 metric tonnes of gold (Gosselin and Dubé, 2005). Realizing that the Witwatersrand nuggets originated far from where they were discovered, prospectors and scientists set out to locate their source or place of formation. They discovered veins of gold associated with quartz and, in lesser quantities, carbonates, tourmaline, and pyrite. This search for gold veins led to the realization that gold ore may contain a wide range of minerals and be present in a variety of different geological environments. The association is not random, and certain minerals have greater affinities for gold than others; namely, the sulphurs, the arsenides, and the tellurides of iron, copper, and other metals.

Subsequent technological advances in the analyses and extraction of gold ore led to the discovery of a previously unknown type of ore, now called invisible gold. This ore is found in very small particles in the structure of minerals such as arsenopyrite (an iron pyrite) and arseniferous pyrite. Ore of this type can mislead the uninitiated; the only way to confirm the presence of gold is by chemical analysis – the gold is so fine that it cannot be seen, even under an optical microscope. A significant portion of present-day gold production in North America comes from huge such ore deposits, especially, the Carlin-type deposits in Nevada, which have made that state one of the most important gold-producing regions in the world. Because of these Carlin-type deposits, the United States is the third-largest gold producer in the world, behind South Africa and Australia.

Gold is also found as a by-product associated with base metals (copper, lead, zinc) and uranium deposits. Deposits that combine base metals and precious metals (gold and silver) are highly sought after for their profitability because the value of this ore is less sensitive to market fluctuations. On the polished surface of an ore sample observed through a microscope in reflected light, the gold stands out for its shiny yellow colour and its great capacity to reflect light.

Gold Deposits in the Americas and Ore Production in the Twentieth and Twenty-first Centuries

The production and gold reserves (or productivity) of the Americas are illustrated in the sidebars of Figure 1. They account for 24 percent of production worldwide. These numbers do not take into account gold production before the twentieth century, or the production from deposits mined using artisanal techniques or by placer extraction (sand and gravel). Figure 1 also shows the geographic distribution of the most important giant gold deposits in the Americas. A gold deposit is said to be "giant" when its past production, combined with its current reserves, totals more than 300 metric tonnes. Giant deposits are found in several types of natural rock of different ages, and there are at least two large gold-bearing districts in the Americas: the

The Carlin site, Nevada
Illustration of the size of deposits exploited.

Photograph: Bettermines

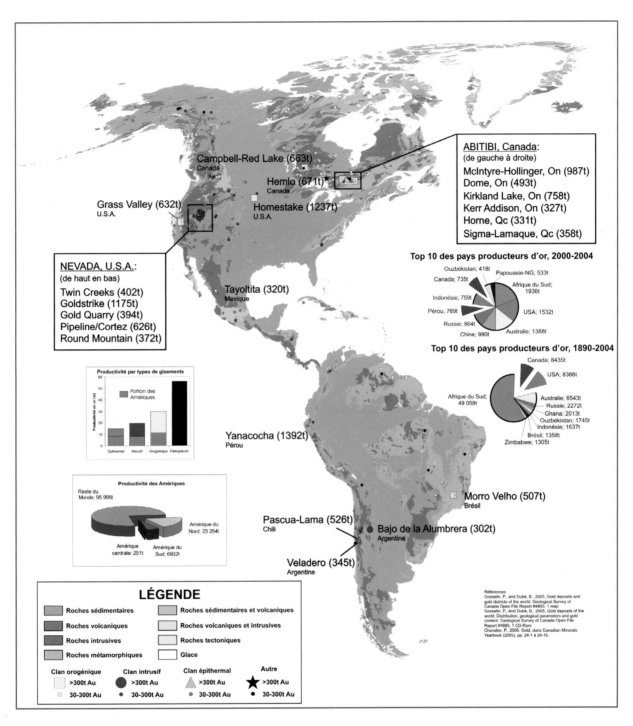

Figure 1

Map showing giant gold deposits in the Americas (>300t Au), 2003

Graphic design: Benoît Dubé and Patrice Gosselin,
Geological Survey of Canada, Natural Resources Canada

Abitibi region, stretching from Timmins, Ontario, to Val-d'Or, Quebec, and the Carlin Trend region in Nevada. Although the geological characteristics of both these regions are well known, scientists still cannot explain exactly why they are so rich in gold. A multitude of geological events ranging in scale from the level of Earth's crust to that of a micron and involving complex physical and chemical processes are the cause.

Several types of gold deposits are known, all formed in different geological environments, at depths that vary from the Earth's surface to more than 10 kilometres (Fig. 2), and in different geological periods, from as far back as 2.7 billion

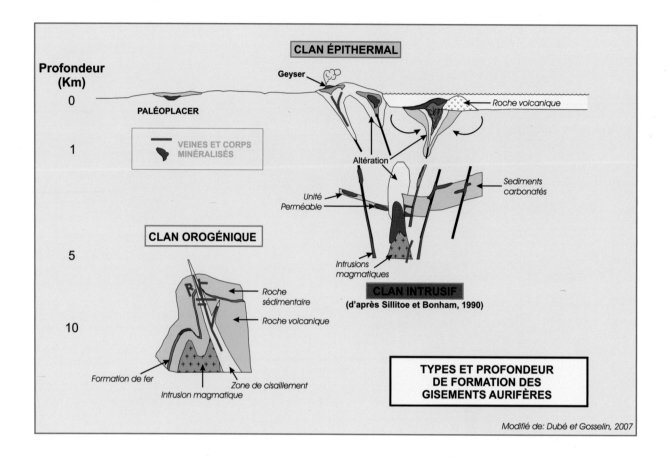

Figure 2

Types and depth of formation of gold deposits, 2007

Graphic design: Benoît Dubé and Patrice Gosselin,
Geological Survey of Canada, Natural Resources Canada

years (in Canada), up to as recently as 10 million years ago (in Chile and Peru). Gold ore is extracted from mine shafts that can descend over 1.5 kilometres (5,000 feet) underground, as at the Campbell-Red Lake Mine in Ontario. It may also be extracted from open pit mines, as is the case in the Carlin site in Nevada and the Yanacocha site in Peru. The productivity and types of deposits in the Americas are illustrated in Figure 1. Note the predominance of orogenic-type deposits (10,960 t Au). The only type of deposit not found in the Americas is the paleoplacer type. The paleoplacer deposits at Witwatersrand in South Africa have a total gold content on the order of 78,000 t Au, which represents about 60 percent of the world's gold (in production and reserves; Gosselin and Dubé, 2005)!

Some deposits, notably those in Chile, Peru, and Argentina, were formed close to Earth's surface during sub-aerial volcanic activities (neutral and acidic epithermal deposits; Fig. 2). Other deposits, formed in environments of sedimentary limestone, are associated with a complex combination of

geological events such as deformations (faults and folds) and magnetic activity, and their formation may have been influenced by climatic changes related to tectonic uplifts of several kilometres, as in the Carlin-type deposits in Nevada (Fig. 2). Most of the gold in Canada comes from deposits formed at depths of between 5 to 10 kilometres under the surface. These deposits are associated with the collision of microplates or assemblages of distinct types of rock during periods of seismic activity and the massive circulation of hydrothermal fluids (orogenic deposits, Fig. 2).

Techniques of Gold Extraction

In the development of every civilization, there is a link between the geographic environment and the exploitation of this environment using simple technologies. The evolution of these technologies takes place in the context of an influx of ideas from

Trapiche (ore-grinding mill), region 3, Chile
Photograph: Marcelo Riveros, 2007

elsewhere, and the evolution of gold extraction is no exception (Kidder, 1964).

The underlying principle of gold extraction is tied to the state of the gold ore in nature, commonly known as the degree of liberation; that is, the ease with which gold may be separated from other minerals associated with it. When gold is free state, the extraction process is less complicated than when it is deeply embedded in other minerals such as quartz or is trapped in minerals such as pyrite, chalcopyrite, arsenopyrite, etcetera, in which case more elaborate extraction techniques are required.

Economic, mineralogical, and environmental factors must also be considered. Ores that are low grade and of simple mineralogy generally require low-cost extraction processes. Ores with a very complex mineralogy, especially the sulphurs, demand very elaborate processes. The choice of process must also take into account the environmental consequences.

Free gold, in the form of nuggets and flakes, is recovered mechanically on the basis of its specific weight, which is higher than that of other associated minerals. The most rudimentary technique based on this principle is panning, in which a large, shallow, round pan is filled with water and the gold-bearing sediment. Using a circular movement, and holding the pan at an incline, the lightest particles are sluiced off while the gold, which is heavier, remains in the bottom of the pan. The same principle of separation by gravity, using water combined with the effect of vibrating tables, is used in more elaborate mechanical systems.

To concentrate free-state gold that is visible to the naked eye in veins of quartz, the ore undergoes an initial grinding to free the gold, and is then treated by gravity separation. At the industrial level, the amalgamation process using mercury is now rarely used, but it remains in use in low-output economies and artisanal recovery operations. The process is based on gold's affinity for forming an amalgam with mercury, which when submitted to evaporation results in pure gold.

In cases in which gold is found closely associated with other sulphurs and arsenides or when, after milling, it is very small in size, the gold is extracted by a process of lixiviation by cyanide, better known as cyanidation. After grinding, the ore pulp is washed with a solution of sodium cyanide in the presence of air. This solution is then treated to remove the gold. The most effective present-day method uses activated carbon: the gold precipitates onto the activated carbon, which is then extracted from the gold content using an acid treatment that enables the carbon to be reused.

The successful use of plants to extract low-grade gold from mill tailings associated with the deposit at Kolar, in India, was recently reported (Isloor, 2005). The use of plants to extract metals from ore is known as phytomining. This method, in which plants act as a hyper-accumulator of gold, was used for the first time in New Zealand in 1998. Similar studies were also done during the exploitation of the Brasilero Deposit in Brazil.

Transformation Processes

In order to eliminate base materials such as copper, zinc, iron, and lead (as well as other impurities), the gold concentrate obtained in the extraction process is first submitted to roasting at temperatures of from 600 to 700°C. This enables the metals to be converted to an oxide state and facilitates their elimination in the slag during the fusion process. The fusion process consists of heating the product of roasting to between 1200 and 1400°C using silica and borax (sodium borate). The silica helps to form slags of the base metals and the other

impurities, while the borax helps lower the fusion point of the silica from 1723°C to about 800°C, depending on the ratio between the silica and the borax. Gold and silver, which are heavier than the slag, accumulate at the bottom of the fusion receptacle. The gold and silver content of the melted concentrate can be has high as 95 percent or more. The result is an impure bar of gold and silver called a dore bar, which may be sold as is or submitted to other treatments.

To purify it, the dore bar may undergo a refining process in order to separate out the precious metals. Two processes are often used to arrive at gold of very high purity (99.99%), namely, the Miller process of chlorination and the Wohlwill process of electrolysis.

In the Miller process, the dore bar is melted and submitted to a precise and controlled treatment using chlorine gas. This results in a concentration of gold of from 99 to 99.5 percent, along with trace metals in the platinum group as well as copper and silver. The Wohlwill electrolysis process involves converting the gold in a chlorine solution by dissolving the gold metal in hydrochloric acid in the presence of chlorine gas. The procedure is very efficient in eliminating metals in the platinum group and producing gold that is 99.99 percent pure.

Environmental Consequences in the Twenty-first Century

The large-scale extraction of a mineral or metal from rock demands procedures that have a major impact on the environment, namely, the transporting and treating of a large volume of rock. The challenge is to find a balance between the exploitation of the natural resource and respect for the environment.

In the past, there was little concern for protecting the environment. These days, however, environmental protection is at the very centre of mining: exploration, mine planning, mining activities, and site restoration. As a consequence, current research focuses on the safe and responsible use of our mineral resources.

A large quantity of sterile crushed and broken rock is produced to extract just one gram of gold. This pulverized rock must be discarded either under or on Earth's surface. As a result of their natural composition, some rock leaches acids and

heavy metals into the ground when exposed to the elements. The cyanidation process by which gold is extracted from crushed ore opens the door to potential leaks and runoff that could contaminate surrounding streams and rivers as well as groundwater. During a recent incident in Romania, 120 tonnes of cyanide were accidentally spilled into a river and this pollution spread all the way to the Danube in Hungary (Seghier, 2005). The use of mercury to separate gold from ore has led to health problems among miners and people living in mining communities. To prevent cyanide leaks, the discarded tailings are covered with a composite coating made up of a clay base and a layer of plastic, usually polyethylene or polyvinyl.

Water is essential to the gold-mining industry, and its use considerably reduces the hydrostatic level in certain mining regions. The use of high-pressure water cannons in the recovery of alluvial gold is detrimental to waterways because of excessive corrosion due to water runoff.

In twenty-first-century mining operations, great efforts are being made to prevent or diminish these effects; in particular, to prevent from the outset acidic runoff from mining. When rock of the type known to cause acidic runoff is encountered, the mine takes action to prevent the problem. With respect to tailings, depending on the climate, a variety of techniques may be used to prevent the oxidation of the piled rock.

One little-known environmental consequence is natural pollution. Toxic metals such as arsenic and thallium come immediately to mind; depending on the type of gold deposit, they are present in concentration and at variable range. Climatic conditions cause the rocks containing these metals to disintegrate and change. The resulting ground formation in a given location, combined with the local hydrological network, dispenses toxic metals into the environment, potentially contaminating the food chain, as was the case in one gold-producing region in China (Xiao et al., 2004). An awareness of the presence of such toxic metals in the gold deposit, in the rock in which it is embedded, and in the ground is necessary in order to evaluate the environmental consequences of any mine and ensure that it is safe and healthy.

As a result of advances both in our understanding of both environmental phenomena and

Gold mine operated by folk miners, region 3, Chile
The gold mine always takes its toll.

Photograph: Marcelo Riveros, 2007

mining technology, there has been considerable improvement in gold mining with respect to environmental consequences. Analyses and assessments of the environmental risks are now de rigueur for all mining operations. And mine sites must be reclaimed when the mine closes. Some have even become lush parks. It is noteworthy that, thanks to globalization, techniques perfected in the developed nations have now been adopted the world over. Methods developed in Quebec are being exported, and imitated. Nonetheless, finding a suitable compromise between the methods of gold extraction and safeguarding the environment remains a huge challenge.

Living Conditions of Modern-day Miners

In 1999 in Peru, 30,000 families of miners working in small-scale operations lived in extreme poverty (Proyecto GAMA-Gestion Ambiental en la Minera Artesanal). Working conditions both inside and outside these mines are very harsh; miners use mercury to extract the gold, putting their health at risk. In 2001, the Minister of Mines in Peru sponsored a law to protect small-scale operations from the vicious circle inherent in emerging, informal economies. For the first time, Peruvian miners could acquire their own concessions or, in the case of existing concessions, sign legal contracts involving them.

For large numbers of miners throughout the world, artisanal mining often goes hand in hand with terrible working conditions. One example is the *garimpeiros* of the Amazon Gold Rush, working their small placer deposits. In Mongolia such miners are called "Ninja Turtles" for the green tubs of ore that they carry on their backs (International Labour Organization, 2005). It is estimated that in

Mongolia there are 100,000 *ninjas*, a group made up of unemployed workers and former herders, who having lost their herds due to devastating winters, were forced to abandon their nomadic lifestyle. The increase in mining activity there, due to the presence of foreign mining companies, necessitated the adoption of new laws to improve the situation in artisanal mines. There has also been an increase of women working in small-scale gold mines. Although women were traditionally prevented from working in the mining industry in Brazil, China, and elsewhere because of sexual discrimination, they are nonetheless part of the workforce in small-scale mines and are thus taking their own livelihoods in hand.

In North America, mining corporations have improved the actual work of mining. This, in conjunction with technological advances, has impacted favourably on workers' health and living conditions. New labour laws related to miners' safety and working conditions have also led to positive change, and mining is now carried out in a much safer environment. The lives of miners living in isolated communities have changed little, however. Towns of different sizes grow up around mining centres (for example, Timmins and Kirkland Lake in Ontario). When new mines open in distant northern regions, miners must adjust their work schedules; they live for long periods at the mine site, returning to their homes farther south for rest and recuperation.

Astronaut Eugene A. Cernan, commander of the
Apollo 17 mission, on the surface of the Moon, December 13, 1972
The visor of his helmet is covered with gold
to better reflect the sunlight.

Photograph: NASA, Inv. AS17-134-20476

Diversity of Industrial Uses of Gold Today

Since time immemorial, gold has been esteemed in jewellery making and in decorating. One of its best-known applications is as gold leaf, produced by manually beating gold until it becomes very thin. Pure gold or gold associated with a low percentage of silver or copper can be beaten into sheets that are 0.0000127 centimetre thick (Boyle, 1979). These sheets are used to decorate buildings, monuments, windows, and even, in India, sweet confections. Thin coats of gold are also used as gilding on a variety of small household objects such as quills and pens, lighters, bathroom faucets, and so on. Liquid gold is used in the decoration of fine ceramics and glassware, using a process by which gold is placed in suspension.

Gold is used in numerous industrial applications because of its natural properties; these usages are described in detail on the Web site of the World Gold Council (www.gold.org). According to Corti and Holliday (2003), only 12 percent of the total demand for gold is generated by industrial applications, compared to 50 percent for platinum, palladium, and silver. With its high thermal and electrical conduction properties in combination with its resistance to corrosion, gold is regarded as a high-performance material in the fabrication of electronic circuits for the telecommunications and information technologies industries. In addition, nano-scale gold particles (10^{-9}) display very interesting properties for the field of optics (op. cit.).

According to the World Gold Council and www.responsiblegold.org, some recent studies have confirmed gold's potential as a catalytic agent in speeded up chemical reactions during which the gold is neither absorbed or consumed. Gold's advantage as a catalyst is the low temperature (200–350°K) at which it becomes functional, compared to that of platinum, which has a optimization temperature of between 400 and 800°K. Certain chemical processes such as hydrogen technology, the control of environmental pollution, and the control of emissions by some types of motors also use gold as a catalyst.

Gold is used in computers, cell phones, the automobile industry, and even the manufacture of CDs and DVDs. Gold also plays a key role in construction and in space programs, largely because of its effectiveness in reflecting heat and infrared radiation. Thin layers of gold cover the windows of many of today's skyscrapers as well as the visors of astronauts' helmets.

The use of sheets of gold to fill dental cavities goes back to before the Common Era. Today, dental work uses mostly gold alloys. Gold is an ingredient in certain drugs and has been used in Ayurvedic treatments in India, Egypt, and China for several centuries. Research is being carried out on the use of gold in pharmaceuticals and medicine for the treatment of various diseases, including cancer. The biomedical industry uses gold in the manufacture of parts for implants and drug delivery microchips because of its resistance to bacterial infection, its biocompatibility, its malleability, and its resistance to corrosion. These microchips may be swallowed, surgically implanted, or administered intravenously. Gold nano particles are also used in biomedical diagnostics (Corti and Holliday, 2003).

Is Gold Irreplaceable?

Several metals could compete with gold, but the usage of any metal is determined by its characteristics. Platinum could be the metal of choice in the manufacture of jewellery; its white colour makes it especially appealing. In the past five years, platinum jewellery has proved a market favourite. White gold (an alloy of gold, copper, silver, palladium, as well as small quantities of nickel and zinc) doesn't have the same hardness, durability, or luminescence as platinum. So why isn't gold being replaced by platinum in jewellery making? Platinum costs more than gold, and the latter is easier to work because its melting point is lower than that of platinum. Platinum will not replace gold as the metal of choice in gilding and decoration, but it will continue to compete with gold as a catalyst in the catalytic converters used in exhaust systems in cars (Lassonde, 1990). Recent statistics have shown platinum's encroachment in this sector. In China alone, the demand for platinum in the automobile industry has risen sharply, from 75,000 ounces in 2004 to 110,000 ounces in 2005 (Ciaccio, 2006).

In the field of computational nanotechnology, studies have been carried out to assess the efficiency of metals other than gold in the production of computer chips. Researchers have, for example, discovered that germanium may be placed on silicon wafers in a more consistent fashion than gold (Pescovitz, 2005). The problem with metals such as germanium is that they do not occur in ore deposits. Germanium is a by-product of zinc refining or is released during the combustion of coal.

In the electronic industry, gold has been replaced by alloys of silver, copper, and palladium. However, with the increase in demand for electronic products, gold consumption in this sector remains at about 4 million ounces annually.

In dentistry, gold faces strong competition from an alloy of platinum and palladium, but especially from ceramics, which are less expensive and look more natural in the mouth. The decline in the use of gold in dentistry has been significant, from 2.4 million ounces in the 1970s to 1.4 million ounces in the late 1980s.

Conclusion

Gold has occupied a central place in people's lives through the centuries because of its properties. The future of gold and of its many uses nonetheless face challenges. The search for new deposits must adapt to the variation of modes of formation as well as the natural distribution of gold ore. Ways must be found to improve the exploitation of these natural habitats and also ore extraction, although environmental protection is now part of the basic equation of the mining industry. But in the end there's no denying that this precious metal has had an enduring influence on us: it seems nothing will ever replace gold.

Bibliography

Boyle, R. W. 1987. *Gold, History and Genesis of Deposits.* New York: Von Nostrand Reinhold Company Inc.

Boyle, R. W. 1979. *The geochemistry of gold and its deposits.* Geological Survey of Canada, Bulletin 280.

Ciaccio, Alison G. 2006. *The Bling Factor: Platinum,* http://www.grandich.com/docs/barrons_06-12-06.pdf.

Corti, C., and R. Holliday. 2003. "A golden future." *Materials World,* February: 12–14.

Gosselin, P., and B. Eubé. 2005. *Gold deposits of the world: Distribution, geological parameters and gold content.* Geological Survey of Canada, Open File 4895.

Isloor, A. M. 2005. "Gold mining with plants?" *Deccan Herald,* 30 August 2005.

Kidder II, A. 1964. "South American High Cultures." In *Prehistoric Man in the New World,* ed. J. J. Jennings and E. Norbeck, 341–86. Chicago: University of Chicago Press.

Lassonde, P. 1990. *The Gold Book.* Toronto: Penguin Books.

International Labour Organization. 2005. *Ruée vers l'or en Mongolie - ou lorsque des bergers se transforment en "ninjas."* Espace Média, http://www.ilo.org/public/french/bureau/inf/features/05/mongolia.htm.

Pescovitz, D. 2005. "Nanoislander: Lab Notes." *Research from the College of Engineering, University of California* 5 (7).

Poulsen, K. H., F. Robert and B. Dubé. 2000. *Geological classification of Canadian gold deposits.* Geological Survey of Canada 540.

Proyecto GAMA-Gestion Ambiental en la Minera Artesanal - Peru (Gestion de l'environnement dans les exploitations minières artisanales au Peru). *All that glitters is not gold.* Swiss Agency for Development and Cooperation, http://www.sdc.admin.ch/en/Home/Projects/Environmental_management_in_Peruvian_gold_mining.

Seghier, C. 2005. "Les déchets miniers seront bientôt encadrés par une nouvelle directive à l'échelle européenne." *Actu Environnement,* http://www.actu-environnement.com/ae/news/1420.php4.

Sillitoe, R. H., and Bonham, H. S. Jr. "Sediment-hosted Gold Deposits: Distal Products of Magmatic-Hydrothermal Systems." Geology, Vol. 18, No. 2 (1990), 157–61.

Xiao, T., J. Guha, D. Boyle, C.-Q. Liu, B. Zheng, J. Chen, G. C. Wilson and A. Rouleau. 2004. "Naturally-occurring thallium: A hidden geo-environmental health hazard?" *Environment International* 30 (4): 501–07.

The *Buisson d'or*

MICHEL GUIRAUD

Professor of mineralogy and head of collections, Muséum national d'histoire naturelle, Paris

Translated by Joan Irving

THE MUSÉUM NATIONAL D'HISTOIRE naturelle began its collection in 1635 with the founding of the Jardin du roi, created to improve man's understanding of nature and to undertake there the most exhaustive possible inventory of nature. Although the scientific approach has clearly dominated in its collections, the institution has remained in some respects a "cabinet of curiosities" throughout its long history. The search for the finest, rarest, biggest specimen – one that reflects not only nature's diversity but also exceptional characteristics as a cultural and aesthetic object, and as a scientific specimen – has always motivated our acquisitions.

The mineralogical collections hold their share of beautiful pieces, including that masterpiece of nature known as the *Buisson d'or* (Golden Bush). In the late 1980s, the collection's curator, Henri-Jean Schubnel, received financial help from the Elf Corporation, the petroleum company now owned by the Total group, to put together a collection of specimens that would be exceptional for their crystallographic quality as well as their mineralogical rarity. His successor, Pierre-Jacques Chiappero, kept this project going, and, with the support of the Total group, the Museum was able to acquire over 300 specimens.

The *Buisson d'or* is one of the highlights of the collection. Acquired in 1993 and listed as No. 194-8, it is a gold specimen on a gangue of white quartz weighing 1.9 kilograms and measuring 27 centimetres high, 22 centimetres across, and 1.5 centimetres wide. Made of an aggregate of plurimillimetric gold crystals, it came from the Eagle's Nest Mine in California, which mines several veins whose sole economic value derives from the crystallized gold pieces found there. These are sold at far higher prices than that for an ingot. At its mine in Placer County, the gold is mined in an open pit and also underground in the form of placers made up of alluvial deposits of Tertiary and Recent age. This mining region is located in the vast metamorphic complex of the Foothills Metamorphic Belt, a formation composed of metamorphized sediments from the Paleozoic era and interposed series of volcanic rocks. These rocks are interspersed with veinlets of quartz about 5 centimetres wide. Crystallized gold is found in some of the veinlets, which are unusual in that they contain very little sulphur or tellurium, minerals frequently associated with gold. The latter displays almost constant purity of from 870 to 880 on 1,000 (that is, 21 karats). Because the mine focuses on finding specimens destined for sale as minerals, the techniques used there are relatively non-intrusive. The gold is prospected using a metal detector and excavated by hand. Each specimen is then worked to remove the quartz gangue so that only the gold is visible. The *Buisson d'or* is representative of the crystalline forms found in the Eagle's Nest Mine, flattened octahedrons tending toward dendritic branches. The thinness of the mine's quartz veins results in specimens that are, at certain sizes, essentially two-dimensional.

Some of the fame of the *Buisson d'or* arises from the origin of its name. Writing in *Trésor du Muséum*,[1] published in 1998, Henri-Jean Schubnel tells the following story. Benoit Mandelbrot, the

◄ **Native gold on white quartz**
The *Buisson d'or*, a specimen that is extraordinary
for its crystallographic quality and mineralogical rarity.

Eagle's Nest Mine, Placer County, California
Muséum National d'Histoire Naturelle de Paris, Inv: MNHN 194.8
H: 27 cm W: 21 cm D: 1.5 cm
Photograph: Alain Dahmane, Optique St-Hilaire

---

1. Henri-Jean Schubnel, *Trésor du Muséum* (Paris: Éditions du Muséum national d'histoire naturelle, 1998).

eminent mathematician, used this specimen to illustrate the cover of his book *Fractales, hasard et finance*.[2] The author speaks of a "bush" or a "small golden branch" that can be seen at the Muséum national d'histoire naturelle. "This artistic masterpiece from nature symbolizes this book because it combines the three themes alluded to in the title … [This bush] is a polysynthetic aggregate generated by innumerable but poorly understood natural forces … The fractality is manifested by the presence of holes and other characteristic structures covering a very wide range of sizes. Only one tool is available to represent the results of these forces, chance … I hope that you find it symbolic that the bushy substance illustrating the cover is almost pure gold."[3]

The cyclical but discontinuous repetition of the same forms and motifs is, as it happens, the characteristic of fractals. The author sees chance in the fact that the physicochemical phenomena which led to this crystallization are numerous and complex, and that forms like that of the *Buisson d'or*, which are frequent in nature, nonetheless defy prediction by statistics. For Mandelbrot, the materiality of this object can be explained only by mathematical chaos. The *Buisson d'or* is one of those slightly magical objects that carries many meanings and therefore transcends science and culture.

2.   Benoit B. Mandelbrot, *Fractales, hasard et finance (1959-1997)* (Paris: Champs Flammarion, 1997).

3.   Ibid., p. 1–2.

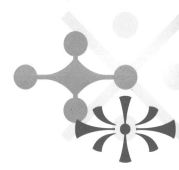

Mineral Exploration in the Twenty-first Century

JEAN-MARC LULIN

Geologist, President and Chief Executive Officer, Azimut Exploration Inc.

THE HUNT FOR GOLD in North America began with the Amerindians thousands of years ago and continued more recently with the arrival of explorers, fur traders, prospectors, and geologists. In the nineteenth century, the systematic exploration of vast territories soon led to the discovery of world-class deposits, some containing more than 150 tonnes of gold (around 5 million ounces). These discoveries catalyzed the economic development of entire regions. Today, the search for new deposits relies heavily on past successes: knowing what has already been discovered, and where and how it was discovered, is a necessary part of developing exploration strategies that will lead to the next economic success.

A Structured Industrial Activity

The mining sector is a field of structured industrial activity involving several stages: exploration, development, mining, and, finally, environmental rehabilitation when reserves are depleted and mining comes to an end. The aim of the exploration stage is to discover new ore deposits, not only to replace mined-out deposits, but also to satisfy an increase in demand. A rise in demand, expressed in a simple way, is reflected by a higher market price for a given metal. This demand is what drives mining companies to look for new deposits in the remotest regions of our planet. But when global demand declines, market prices slump and the reach of exploration activities is pulled back, with

◄ Boomer XE3

This drilling rig is a high-performance piece of equipment.

Atlas Copco, Marketing Communications URE, #205605
Photograph: Goran Wink

companies tending to look to ore deposits already in production.

Several strategies emerge from this global dynamic: first, stay within established mining camps or mining regions where the potential has been proven; second, move into less explored regions without an established potential but presenting a similar geological setting to producing regions; third, open up new frontiers in territories with poorly known or unknown geological settings and an uncertain potential.

From Mining Camps to New Territories

The companies' strategic choices about exploration correspond to a risk analysis. In a mining region with established infrastructure and proven potential, the risk seems lower. However, the chances of discovering a new ore deposit in these regions diminish over time because decades of activity mean that most significant discoveries have already been made. At this stage, the mining camp is considered mature. In contrast, the risk may seem higher for regions that have seen little or no exploration, but the prospect of finding a major ore body at the surface and being part of a new cycle of discoveries is very attractive. This question of "mining camp or new territory?" must be contemplated by every company exploring for gold.

Twentieth-century exploration in Canada led to the emergence of world-renowned gold mining camps such as Val d'Or, Timmins, Kirkland Lake, and Red Lake. Today's challenge in these regions is to renew reserves and keep up production levels. The easiest deposits to find are those that crop out at the surface. Over time, it makes sense that the mineral potential is developed deeper along the extension or in the vicinity of known deposits. Producers strive to realize this potential

Aerial view of the camp for Project Éléonore,
James Bay sector, Quebec

Photograph: Mines Opinaca

until all hope of efficiently renewing their reserves must be abandoned. For this reason, the example of Osisko Exploration Ltd. in the Malartic mining camp of the Abitibi region is particularly remarkable: in 2005, the company "rediscovered" the deserted Canadian Malartic deposit based on a new exploration model with the goal of opening an open pit mine rather than an underground operation. Osisko has since declared estimated resources of 250 tonnes of gold.

The decision to move into new territories has been made by many companies, including active producers, known as "majors," and non-producing companies, known as "juniors." The common perception is that it is easier to make big discoveries, and at lower cost, in poorly or under-explored regions. Nevertheless, the cost of the infrastructure required to put the first mine into production can be very high: sometimes several hundred million dollars to more than a billion. For these reasons, the first discovery in virgin territory must be large enough to justify the construction of the necessary infrastructure. The discovery of such a deposit, called a "founding deposit," leads to the establishment of a sustainable regional dynamic in which two components dominate: a) competition between companies to acquire and explore targets similar to that of the initial discovery; and b) leverage created by the new infrastructure that makes it easier for subsequent discoveries to be economically viable.

In Canada, exploration is migrating northward, where an increasing number of analogues to the geological settings of developed mining regions are being recognized. In Quebec, it was the James Bay region that grabbed the attention of a number of companies back in the 1990s. This vision eventually led a pioneering junior company, Virginia Gold Mines Inc., to discover the Eleonore deposit in 2004, for which resources are currently estimated at 90 tonnes of gold. Goldcorp Inc., a major company, has since acquired the deposit and foresees a potential of 150 tonnes with production scheduled to begin in 2010. The discovery of Eleonore started a real boom in the James Bay region, quickly attracting more than 50 exploration companies and new capital. These conditions make additional discoveries much more likely, and could lead to the creation of a new and permanent mining region.

Our planet is truly opening up to mineral exploration. Territories that were once considered inaccessible or too risky are today's destination of choice for many geologists. And the past 20 years have seen Canadian companies play an important role in expanding exploration around the world, particularly for gold. The phenomenal West African exploration boom began in the early 1990s with the discovery of the Sadiola deposit in Mali by the Canadian company IAMGOLD. The same effect occurred in 2006, when another Canadian company, Aurelian Resources Inc., discovered the enormous Fruta del Norte deposit in Ecuador, announcing an initial estimate of 425 tonnes of gold, but maybe with a potential for 500 tonnes or more! This expansion of exploration activities, accompanied by the perception of diminishing socio-political risk, is encouraged by many countries hoping to attract mining investments.

Science and Some Luck

There are no miracle methods for discovering a gold deposit. The approach relies on the description of previously discovered deposits and their geological settings to identify analogue environments. The knowledge base for gold deposits must evolve from the scale of the deposit (and of the mining camp) to the continental scale before geologists can truly understand the factors favouring or controlling the emplacement of potentially economic concentrations. Different types of gold deposits (epithermal, porphyritic, orogenic, etc.) correspond to specific geological environments, and so if new deposit types are identified, it prompts companies to explore settings for which the potential had never before been recognized.

The mining industry has on hand a panoply of methods to explore geological environments that it considers promising. The choice of which methods to use will depend on the exploration history of the target region (mature camp or virgin territory?), the geographic location (northern or tropical?) and the type of deposit sought. In addition to acquiring a knowledge base, such as geological mapping or remote sensing data, the main approaches include surface prospecting; geochemical surveys that detect gold and other accompanying elements in geological rock formations and also the secondary surficial environments (soil or stream and lake sediments); geophysical surveys that detect favourable geological formations and minerals typically found with gold by characterizing their physical properties, such as magnetism, conductivity, resistivity, and chargeability; and drilling, a crucial stage at which the information obtained from drill cores can be used to determine, by interpolation, the volume of mineralized rock. If the calculated volume is of sufficient grade to profitably extract the metal it contains, then voilà! An ore deposit has been found!

The discovery of a deposit is almost always the combined success of many methods. Targets identified by prospecting, geochemistry or geophysics must be tested by drilling. Paradoxically, the great variety of methods and the abundance of available data do not necessarily translate to greater success in exploration. One of the challenges for explorationists today is to increase the quality of the initial targeting process so that selected targets truly have better odds to correspond to ore deposits. Data can be processed by digital methods, which may lead to develop predictive mineral potential modelling, although the industry's use of such models is still in its infancy.

The inherent risk in exploration can be reduced by a combination of regional geoscientific data, knowledge of deposit types, and practical know-how. The possibility of new major gold discoveries is undoubtedly considerable, but it is never a rapid process or an easy task! And even if the use of scientific methods increases the probability of success, luck remains an inseparable component of discovery that can never be completely eliminated. Luck is the unpredictable friend of explorationists. To rely on it would be disastrous, but it will always be there ... even in the twenty-first century!

Open-pit gold mine,
Sadiola, Mali

Photograph: IAMGOLD Corporation

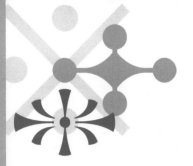

Agnico-Eagle Mines Limited

EBERHARD SCHERKUS

P.Eng., President and Chief Operating Officer

Translated by Joan Irving

Illuminated
mine roadway

Agnico-Eagle Mines Limited
Photograph: Serge Gosselin

AGNICO-EAGLE MINES LIMITED, a mining company with more than thirty-five years' experience in the production of precious metals, is famous for the dynamism of its exploration activities. The company is particularly active in the Abitibi-Témiscamingue region, where it operates the Laronde Mine, near Cadillac. With its more than 1,000 employees, Agnico-Eagle is very active in the socio-economic development of this region. It aims to build a company focused on quality, growth, and respect for its employees and the environment. The company is, in addition, on a very solid financial footing.

History of the Laronde Division

Founded in 1961, the Laronde Division was originally called the Dumagami Mine. Preliminary exploration work was conducted in 1963 and 1964, but was suspended until 1974, when a surface exploration program was begun. This too was suspended, because of low gold prices. In 1980, a surface extraction program was initiated; it was followed in 1983 by an underground program involving the digging of a 1,420-foot shaft. The results of this drilling program were, however, disappointing, and a new approach was recommended. A section of the property located west of the known mineral zones was explored. This drilling program was a huge success and spurred the company's board of directors to authorize a mining operation at the site.

The property went into production in July 1988, at a production rate of 1,500 tons a day. Several modifications and investments have been implemented since then. Currently, the rate of production is 8,000 tons a day, and exploration is going on at a depth of 7,000 feet. Gold, silver, copper, and zinc are being extracted.

Given its past success, the management of the company decided to invest massively in the mine's future development. More than $350 million is being invested to extend the shaft to a depth of 10,000 feet, which will make it one of the deepest mines in the world. This development project will be completed in 2010–11 and will prolong mining at this site for another fifteen years.

The Éléonore Project: Creating the Future!

CLAUDE-FRANÇOIS LEMASSON
P. Eng., General Manager – Projects, Canada & USA

Aerial view of a gold-prospecting exploration camp in Project Éléonore

Goldcorp Canada Ltd.
Photograph: IDNR-TV Natural Resources Television

THE ÉLÉONORE PROPERTY IS located in the northeastern part of the Opinaca reservoir, approximately 190 km east of the Cree community of Wemindji and 320 km north of the city of Matagami, in the James Bay region of Quebec. The geological formations at Éléonore are Archean, being roughly 2.7 billion years old.

In 2001, Virginia Gold Mines' regional reconnaissance work in the area led to the "re-discovery" and evaluation of Noranda's Ell Lake showing. The 2002–03 exploration program led to the discovery of the gold-bearing boulders glacial train in an outcrop located some 6 km northeast. This outcrop became the Roberto deposit.

Goldcorp acquired the Éléonore gold property from Virginia at the end of March 2006, and then incorporated Les Mines Opinaca Ltée and began a drilling program aimed at better defining the known mineralization and its continuity. The holes drilled in 2006 (245 diamond-drill holes, for some 76 km of drill core) reported many assays greater than 10 grams of gold per tonne.

Project work that will be completed in 2008 includes an advanced exploration program, an impact assessment, advanced environmental studies, geotechnical work, metallurgical studies, a feasibility study, permanent infrastructure design, and construction of an airstrip and a permanent access road.

The philosophy of sustainable development has been an integral part of our project ever since Goldcorp-Opinaca acquired the project; our main sustainable development objective is to ensure that there is a balance of economic prosperity, environmental responsibility, and community support in order to develop a safe, profitable, and sustainable mining operation.

Money, money
By Fred Ebb, from the film Cabaret (1972) and performed by Liza Minelli

Money makes the world go around
the world go around
the world go around.
Money makes the world go around
It makes the world go 'round.

A mark, a yen, a buck or a pound
a buck or a pound
a buck or a pound.
Is all that makes the world go around
That clinking, clanking sound
Can make the world go 'round

Money money money money
Money money money money
Money money money

CHAPTER **8**

GOLD, POWER, AND COMMERCE

The Financial Role of Gold: From the Beginnings to the Present
Jean-Bernard Guyon, Analyst, Natural Resources, and Fund Manager,
Global Gold and Precious, for Global Gestion France

The Royal Canadian Mint
Ian E. Bennett, President and CEO, Royal Canadian Mint

The Financial Role of Gold: From the Beginnings to the Present

JEAN-BERNARD GUYON[1]

Analyst, Natural Resources, and Fund Manager, Global Gold and Precious, for Global Gestion France

Translated by Käthe Roth

Origin and Early Days

THE FINANCIAL ROLE OF gold grew gradually over the course of history. With its brilliance and almost total inalterability, gold has become a store of value and assumed a fundamental structuring role in the world economy.

The first role of gold was that of treasure. Gold (and silver) in all its forms, but mainly in the form of goldsmithed pieces, made by artists who had mastered the metallurgy and shaping of gold, constituted the treasures of some civilizations. As symbols of power, these treasures were often the object of covetousness; victorious conquerors quickly stripped the vanquished of their precious ornaments (bracelets, necklaces, inlaid weapons, and so on). What was appropriated from these treasures became "legal tender" over the centuries.

Around 330 BCE, Alexander the Great seized Persia's treasure. The result was a massive influx of gold into Greece. This sudden arrival of "means of payment," given an inelastic supply of goods, triggered a major wave of inflation, upsetting the country's economy. This is the first episode of inflation that historians have found that responds to the definition given by many contemporary economists, for whom inflation is solely a monetary phenomenon.

Gold took on a monetary role very early. At the end of the fourth millennium BCE, an Egyptian king assigned a legal value to gold, materialized by the production of small ingots weighing about 14 grams. The ratio of the value of silver to that of gold, a notion with which modern financiers are familiar, appeared about a thousand years later. This very fluctuating relationship changed largely as a function of the geographic availability of each respective metal. Over the centuries, the gold–silver price ratio seems to have evolved from 2 to 1 to 13 to 1.

Silver, in the form of ingots, was the first metal to be used for trade purposes, 4,000 years ago. Subsequently, precious metals tended to be used widely as a store of value or as a basis for trade contracts in Egypt and Mesopotamia. They then acquired the status of a hedge against inflation, leading directly to the notion of money.

Persia imposed a tax to be paid in gold, while Alexander the Great used the gold seized during his conquests as currency in his empire. The importance of gold was finally accepted when mercenaries and soldiers engaged for military operations began to be paid in this metal. The first gold coins were struck under the aegis of Croesus, the king of Lydia, from flakes of metal gathered from the sands of the Pactol River (ca. 600 BCE). These coins were also employed by the Greeks, who spread their use.

In the first century CE, Rome struck the aureus, weighing 7 grams (gold and silver), and other pieces of lesser value, from the silver denier (2 grams) to the copper-based quadrant. In the fourth century, the solidus (4.4 grams) was put into circulation. Under the Republic and the Empire, use of coins became generalized and millions of coins were struck.

◀ Gold ingots and coins

Photograph: Stockbyte/Getty Images

1. I would like to thank everyone who, throughout my career, has enriched my knowledge with their comments, lectures, publications, and organization of conferences that I was able to attend. It is impossible for me to name them all individually, but they have my most sincere gratitude.

Gold ducats from the shipwreck of the *Girona*

Photograph: Marc Jasinski

The Persians used the silver dirham. Around 700, Caliph Abd al-Malik of Iraq instituted a reform that made the dinar official. With the rise to power of Islamic civilization, the dinar became the currency of reference for a large part of the world, extending from the west coast of Africa to southern Europe and to northern India. It was to be used for centuries and dominated the market until the late thirteenth century.

Starting with the reign of Charlemagne (800 CE), no more gold coins were struck in continental Europe because of the depletion of the European gold mines. In the Middle Ages, gold still circulated, but there was a monetary and financial regression due to a partial return to the barter system.

In the late thirteenth century, Venice put into circulation its first gold coins, ducats. Gradually, they became the dominant European currency, and they remained so for a number of centuries.

Around 1500, several hundred Spaniards accomplished the incredible feat of conquering the Aztec and Inca empires occupying a large part of today's Latin America. This conquest occasioned a massive transfer of gold to Europe after American peoples were stripped of their treasures, composed mainly of worked gold.

The first consequence of the influx of such great quantities of gold into Spain, then Europe, was to provoke a very strong price rise. Spain, however, was not greatly enriched. For one thing, the Jews and Moslems expelled from the country had been an indispensable elite whose absence was sharply felt. For another, engaging in continual and very costly wars helped to create a precarious financial situation.

At the time, the Spanish unit of currency was the escudo. This Spanish coin, struck in great numbers, was as widespread in North America as was English currency and was traded mainly on the Philadelphia market.[2] Around 1700, the idea arose that a more rational organization of the monetary system was necessary in the context of the evolution of economies.

After the collapse of the Western Roman Empire (transferred from Rome to Constantinople in 330), the Byzantines adopted the Roman coinage and retained the solidus as the monetary standard until about 950. After that, until the Byzantine economy and monetary system collapsed, other coins were struck, but their gold content dropped (to only 15% to 30%). According to historians of economics, this could be considered the first time that a currency was devalued.

In the seventh century, trade in the Arab world was conducted in dinars (coins identical to the Byzantine solidus in size and weight), which circulated jointly with the Byzantine currency.

---

2.   The Philadelphia Stock Exchange was founded in 1790.

The Gold Standard

The gestation and institution of the gold standard took place over a fairly long period, and historians of economics still disagree when the first gold standard system was adopted. In 1661, King Charles II of England decreed, through an Order in Council, an innovation that seems obvious today, but was revolutionary for the time: the use of machines to strike coins.

At the time, the coins in circulation were of poor quality, and many were trimmed or tampered with. In December 1695, the king issued a series of proclamations designed to impede the circulation of defective coins. This operation, spread over three years, entailed serious disturbances in the economic system. In 1717, the gold guinea became the coin of reference.

Silver coins had a higher value abroad, which led to massive exports; thus, the Mint, between 1695 and 1740, struck only 1 million pounds' worth of silver coins, as opposed to 17 million pounds' worth of gold coins. In fact, it could be concluded that by 1717 the market in Great Britain had sanctioned gold as a standard, providing it with incontestable supremacy.

The events of the late eighteenth and early nineteenth centuries triggered the global adoption of the gold standard. On February 22, 1797, several French navy frigates raided a fishing port in Wales. This incident caused a panic in England and a gold rush on the banking system, as holders of bank notes (those of the Bank of England and smaller banks) were anxious about the future value of their paper money in case of invasion. A few days after this incident, banks in Newcastle had to close their teller windows due to rumours of invasion. Because of the war against France in 1793, the Bank of England's gold reserves had already dropped sharply, from 7 million pounds at the end of 1794 to 2 million at the end of 1796. It was the first time that the markets attacked a paper currency that was freely convertible into gold (a right that had existed in England for a century). The Order in Council of February 26 and the ensuing *Restriction Act* caused a great shock: a resolution was adopted under which no more gold currency would be issued. Once the acute phase of the crisis had passed, transactions resumed with non-guaranteed bank notes, since everyone believed that there would be an imminent return to a situation

of convertibility. In fact, the embargo lasted more than twenty-four years, until 1821. One problem with the monetary system of the time was the low production of gold in comparison to the development of economic activity. During the eighteenth century, gold production never rose above 20 tons per year; under these conditions, it was normal for government currency and private currency to circulate jointly, in various forms.

One of the major consequences of this situation, and of the development of private credit, was the rise of the private sector as a predominant engine of currency creation. This structure, with a well-established relationship between the money supply and bank credit, was very similar to what we see today. It was not planned at all, and its spontaneous development was an important step in the world's monetary history.

Events moved quickly after enactment of the 1819 *Act for the Resumption of Cash Payments*, followed by the lifting of the embargo in 1821. In 1833, the notes of the Bank of England became legal tender. The 1844 *Bank Charter Act* stipulated that the Bank of England notes were the only legal tender and that they had to be guaranteed by complete coverage in gold. Thus, 1844 marked the establishment of the gold standard in Great Britain.

The crisis hitting paper currency, which was issued in quantity in Europe, was to lead numerous nations to adopt, in turn, the gold-standard system. For instance, after the Franco-Prussian war, Germany created the reichsmark and adopted a strict gold standard in 1871, a decision made easier by the increase in its store of gold from the mines of South Africa. Germany was quickly followed by the Latin Union (Belgium, Italy, Switzerland, France) in 1873, Scandinavia (Denmark, Norway, Sweden), and the Netherlands in 1875, France (outside of the Latin Union) and Spain in 1876, Austria and Greece in 1879, Russia in 1893, Japan in 1897, India in 1898, and the United States (*de jure* in 1900). By the end of the century, only China was still using a silver standard.

From then on, bank notes, foreign currency, and bank deposits owed their acceptability solely to their convertibility into gold. The system that had just been created was the first international monetary system defining the nature of the reserves used to regulate international exchanges and terms for fixing the prices of currencies. It also sanctioned the triumph of gold over silver. Implementation of the gold standard was encouraged, notably, by the increase in gold production in the nineteenth century, a consequence of discoveries and gold rushes in Australia, California, and the Klondike-Yukon.

Problems with the Gold Standard: from 1914 to the Bretton Woods Agreement

The stability created by the adoption of the gold standard was thrown into jeopardy when the First World War started. This conflict forced Great Britain to abandon convertibility of bank notes, and the nations concerned printed more and more currency, leading to sudden and serious inflation. The United States renounced convertibility in its turn. After the hostilities, the Treaty of Versailles imposed major war reparations on Germany and its allies. Therefore, Germany was stripped of its gold and experienced an unprecedented wave of inflation in 1923–24 (prices were multiplied by 500 million!).

A return to the gold standard was considered necessary for prosperity. The Geneva Conference, convoked on the initiative of Great Britain in 1922, brought together all of the countries that had participated in the war, except the United States. This meeting resulted in an agreement setting out a return to convertibility as quickly as possible, in spite of a radically changed global geopolitical situation.

The return to convertibility took place at pre-war parity, an absurd decision that nevertheless received more or less general agreement. The loudest objection was made by the English economist John Maynard Keynes, who predicted a collapse of the economy and called the gold standard a "barbarous relic," a retort that remains sadly famous. The 1925 *Gold Standard Act* did not completely re-establish the previous system. The Bank of England notes in circulation were no longer convertible into gold coins, but only into gold ingots weighing about 12 kg, which limited the possibility of returning to convertibility. In late 1926, France returned de facto to the gold standard, but with a franc worth only 20% of its pre-war value.

In 1928–29, the rise in interest rates in the United States exerted pressure on the Bank of England's gold stores. As gold did not play any role in the stock-market crash of 1929–30, it therefore inspired renewed respect. Bank credit was not available, and so the price of consumer goods collapsed, provoking a series of bankruptcies of banks and businesses. Compared to 1929 figures, prices plummeted by more than 30% and unemployment rose to 25%. Such an economic situation strikes us today as totally aberrant, but it corresponded to the rules known at the time.

Because it had not had any influence in 1929, gold made a return to power on the economic scene. During the years that followed, in an increasingly chaotic situation, all countries had one objective: to protect their gold stores. The shock wave propagated, and the monetary system was shattered. The forty-seven nations that had opted for the gold standard began, with more or less haste, to detach themselves from it. The situation in the United States deteriorated so badly that wholesale prices at their lowest were, in 1932, 40% lower than in 1929. In England, each deflationary measure brought a bigger deficit the following year. Now, the situation was ripe for a radical change that could not avoid involving gold.

12 kg gold ingots

Photograph: Stockbyte/Getty Images

In 1932, the Democratic candidate, Franklin D. Roosevelt, swept into power with a crushing electoral defeat of the Republican, Herbert Hoover, and big changes were in the wind. On April 5, 1933, the new president signed a decree requiring everyone holding gold to take it to the bank and exchange it for paper currency, and the banks had to turn the gold over to the Federal Reserve Bank. Another statute, signed the same month, gave the president the power to fix the weight of gold purchased by a dollar at between 50% and 60% of the weight established in 1837. On June 5, all of the contractual clauses of gold guarantees were abrogated.

The path was now open to an official devaluation of the dollar vis-à-vis gold. On January 30, 1934, the price of an ounce of gold was set at $35, an increase of 69% over the price of $20.67 set in 1900, which had been in force since 1837. Holding gold was now forbidden. This measure, along with others, led to a recovery of industrial production, which grew by 60% between 1933 and 1937. The rise of gold was confirmed in other countries as well, so that by the mid-1930s one ounce purchased almost twice as much goods and services as it had the previous year.

Anticipation of the rise in the price of gold against all currencies and its effective institution had two consequences. The first one, predictably, was the rise in production of the metal, which went from 750 metric tons per year in 1932 to 1,100 metric tons in 1938, so that the gold monetary reserves now covered 100% of the money in circulation, which was excessive. The second, unexpected, was that, with so much gold available, the best solution was to sell it to the United States, which wanted to buy all available gold at $35 an ounce.

By the time the Second World War began, 80% of the world gold monetary reserves had found refuge in the United States and represented a metal stock of 18,000 metric tons. Once again, war was to change the game profoundly. A full year before the hostilities in Europe ended, in July 1944, delegates from 44 countries met at Bretton Woods, in New England, to develop the new economic and monetary system that would dominate after the war.

The measures adopted owed much to John Maynard Keynes, representing the British Treasury (known for his longstanding opposition to the gold-standard plan) and Harry White, representing the U.S. Treasury. The United States, then at the peak of its power, came out of the war almost intact and held 75% to 80% of the world gold monetary stock. The system formulated at Bretton Woods restricted gold convertibility to a single currency: the U.S. dollar. Dollars could be converted into gold at $35 an ounce, but this privilege was reserved for central banks and similar organizations. The value of the other currencies was defined in relation to the dollar at fixed parities, though adjustments were possible. The new system thus led to creation of the mechanism called the Gold Exchange Standard, which, for most countries, succeeded the Gold Bullion Standard.

The establishment of currency parity in relation to the dollar proved difficult in a world in which there were rules controlling the movement of capital except in Canada, the United States, and Switzerland. The system was designed to encourage the return of countries such as Japan, Germany, and Italy into the camp of democratic nations. The American aid provided by the Marshall Plan resulted in the payment of $13 billion to seventeen countries, including France, Great Britain, Italy, and Germany, between 1947 and 1952. The Bretton Woods system finally came to the end of its useful life; it collapsed in 1971, with other consequences for gold.

Up to the End of the Bretton Woods System (August 1971)

The general provisions of the Bretton Woods system were reinforced by the creation of a new institution, the International Monetary Fund (IMF), whose role was to make short-term loans to countries or central banks incapable of finding financing on the markets or from the authorized sources. The IMF's resources came from contributions by each member state: 75% in its respective currency and 25% in gold, with a few exceptions for the countries that did not have the means. In fact, the new system was based on the hypothesis that the dollar was "as good as gold," a belief that would not be questioned for a good ten years.

The crushing superiority of the United States after the war could not last, and the playing field was quickly levelled by the spectacular recovery of Europe and Japan, which had been the objective of the builders of the new system. Thus, there were increases in both American imports and foreign investment in countries that had become attractive once again. The United States' foreign debt grew and its gold reserves began to shrink. At the same time as it was dispensing international economic aid, the United States launched major domestic social programs, deployed its armed forces all over the world, waged two wars in Asia (Korea and Vietnam), and faced growing military expenditures caused by the cold war with Russia.

As a consequence, the United States' gold monetary reserve melted away, representing a counter-value of only $19 billion in 1960, while the nation's foreign receivables (deposits, Treasury bills and notes, etc.) rose to $20 billion over the same period. Conversion of these receivables into gold would have led to the complete disappearance of the American gold store.

When the price of gold on the London open market reached $40 per ounce, the stage was set for a crisis in the dollar, a phenomenon that was not unpredictable in the sense that it was obvious that the enormous imbalance in favour of the United States at the end of the Second World War could not last. Also, it was much easier to transfer dollars from a holder to an acquirer than to transfer gold – not to mention the advantage of accumulating interest. The United States therefore became the world's banker, as England had been before the First World War.

The only real peril for the dollar seemed to be inflation. The first hints arose in 1956–58, when the U.S. economy suffered a serious slow-down combined with a 7% unemployment rate. However, from 1959 to 1964, the pace of inflation dropped to an average of 1.4% as wage demands moderated, while growth rose to 5%. Although the news about inflation was reassuring, the current-account balance of payment deteriorated greatly, leading to accelerated gold departures during the first part of the 1960s. The American gold store thus fell again, in spite of credit agreements aiming to limit the use of this store by foreign countries,

while taxpayers and companies were asked to limit price and salary hikes through "voluntary" agreements.

In thirteen years, from 1958 to 1971, the store of gold in the United States dropped from $19 billion to $10 billion, while American foreign receivables grew by $60 billion. During the 1960s, European demand for gold seemed to be limitless. France, on the sage advice of economist J. Rueff, converted a large part of its foreign exchange reserve into gold.

In 1961, the United States, Great Britain, and a number of countries in continental Europe formed the "gold pool," designed to stabilize the price of gold at $35 an ounce. This move encouraged investors and speculators to purchase gold. France withdrew from the pool in 1967, and on March 17, 1968, the decision was made to close the pool, which was a failure. In practical terms, official transactions among central banks continued to take place at $35, while private operators executed their transactions on the open market, on which prices fluctuated according to supply and demand. It then became evident that the devaluation of the dollar was only a question of time, beyond the expedients used to try to maintain the parity of the U.S. currency. The central banks of the "defunct" pool agreed to stop converting their dollars into gold as long as the Vietnam War continued, and they also refrained from purchasing gold at $35 to resell on the open market at higher prices.

After 1968, inflationary pressures intensified in the United States, and, given the urgency of decisions to be made, U.S. president Richard Nixon appointed the energetic John Connally Secretary of the Treasury. Connally devised a two-part plan: abandonment of the convertibility of the dollar into gold and a series of price- and wage-control measures to combat inflation. All that remained was to set the date when these would come into effect.

On August 9, 1971, a representative of the British Treasury asked the American Treasury for $3 billion in gold. On August 13, a secret meeting was held at Camp David with the main crafters of American economic policy. On Sunday, August 15, before the markets opened the next morning, the elimination of the convertibility of the dollar into gold was announced (the "gold window closing"). From then on, gold was free to "float" on the market, and the fate of gold's monetary role was sealed.

Hand-stamping the logo of the Royal Canadian
Mint logo on 99.99% pure gold bullion bars
of 400 troy ounces

22.5 cm x 7 cm x 4.4 cm
Photograph: © Royal Canadian Mint –
all rights reserved

The End of the Monetary Role of Gold

There was then a period of exceptional events that affected the price of gold. The breaking of the link between the dollar and gold first created a disruptive situation outside of the United States due to the well-founded fear that the dollar would depreciate. Most exchange markets closed for several days. Japan absorbed $4 billion of losses in two weeks; and the yen appreciated when currency transactions resumed, as did all currencies when the other markets reopened.

A meeting in December 1971, held at the Smithsonian Institute in Washington, D.C., resulted in decisions intended to return order to the financial markets: the fixing of monetary parities ratifying a depreciation of the dollar, the elimination of the 10% surtax on imports, and the setting of a new official price for gold at $38 per ounce. In the end, it had to be admitted that the attempt to maintain fixed exchange rates among the main currencies was a failure. Moreover, in October 1973, the members of the Organization of Petroleum Exporting Countries (OPEC) decided to cut back production in order to raise the price of a barrel of oil from $2.11 to $10. In this inflationary context, the central banks made a strange decision, reflecting their confusion. In November, they went back on their 1968 decision to abstain from purchasing and selling gold, except among themselves, and began to intervene on the open market in London.

In 1975, the holding of gold was once again allowed in the United States, and a market was created for it (similar to that for other raw materials) on the COMEX (Chicago Mercantile Exchange). Now, gold was just one raw material among others.

Thus, in the first quarter of 1975, then in 1978–79, the U.S. Treasury sold 6% of the country's gold store. In the summer of 1975, the IMF abolished the official price of gold and decided to sell part of its gold assets, to the profit of developing countries, with a view to reimbursing part of the contributions of founding members.

The pace of inflation remained high, and these transfers made speculators happy. The price of an ounce of gold went from $64, in late 1972, to $130, between 1974 and 1977. After OPEC raised the wholesale price of oil to $30 a barrel in 1978, the price of gold rose to $500, and it reached a high point of $850 in January 1980. The wildest dreams of all speculators had come true: in 10 years, the price of gold rose from $35 to $850 an ounce, a phenomenal rise in real terms, even taking account of an annual inflation rate of about 12% by the end of 1979.

This extraordinary situation was to lead to a draconian decision. Fearing a serious economic crisis, the new president of the Federal Reserve

One-ounce 99.99% pure gold Maple Leaf bullion coins
on a Hordern, Malon, Edwards press, Ottawa

Bank, Paul Volcker, decided, in the last quarter of 1979, to no longer control interest rates but only the money supply, which would allow short-term interest rates to range freely.

Volcker's policy had the consequence of raising short-term rates to extremely high levels, which put a brake on the pace of long-term inflation; this position was consolidated by Volcker's successors. Under these conditions, investing in gold was no longer interesting and the price began to fall. On January 22, 1980, it dropped by $143 in a single session. Between mid-1999 and the first quarter of 2001, gold was selling at $250 an ounce. The drop in purchasing power was 85% over this 20-year period. The decline was accompanied, or accentuated, by sales of gold by central banks starting in 1992, dropping its price between $350 and $400. Between 1992 and 1999, central banks sold about 400 tons of metal per year. The depression of the gold market was further exacerbated in 1997, when Switzerland decided to sell half of its gold store, following the elimination of the provision in the Swiss constitution that currency circulation had to be completely guaranteed by coverage in gold. This created a powerful shock wave. In 1999, the Bank of England followed suit, selling 415 of its store of 715 tons.

When the European Central Bank and the euro were created in 1998, it was provided that 15% of the bank's reserves were to be held in gold, but this did not change the market trend. In total contrast with the situation for the gold market existing in 1980, the market in 1999 had become a cause for concern, and in September central banks in Europe agreed to limit their sales to 400 metric tons per year, with the agreement of the IMF and the U.S. Treasury. The idea was to maintain the value of the central banks' reserves, while for a certain number of developing countries, gold production was a considerable addition to their gross national product.

Another element in the return to equilibrium fell into place. The rise in standards of living and the relative drop in the price of gold compared to other goods led to growth in the jewellery industry that outstripped the increase in mining production. The price of gold began to rise again in the first quarter of 2001, in an economic context that had begun to deteriorate with the bursting of the "Internet bubble" and in a very tense geopolitical context following the attack on the World Trade Center. The general upward trend in prices of raw materials, the weak dollar, a laisser-faire monetary policy, and strong economic growth resulted in a strong gold market in the ensuing years. In addition, the central banks' restraint with regard to their transfer policy was reinforced.

Today, the primary market for gold (mining production + recycling – jewellery + industry) is in a surplus situation, and the rising trend in the market is determined by the combination of investment and speculation, as it has often been in the past.

Use of Gold in Contemporary Finance

After the COMEX gold market was created in 1975, each actor in the market developed, according to need, an activity through the intermediary of techniques or products calling largely upon mathematical knowledge and the advent of computers. The actors in the market were mining companies, central banks, industries, investors, and speculators. Mining companies were not simply producers exploiting their deposits and selling their production. They modulated their sales as a function of either circumstances imposed upon them or decisions that they deemed prudent.

In the second half of the 1990s, it became difficult to find financing to put deposits into production as banks were reluctant to make loans for this purpose; while the trend in metal prices was downward, financial risks were added to the usual operational risks. Some companies proceeded with futures contracts with various terms, thus committing part of their future production. This operation was technically feasible and advantageous as long as the gold price was established at the cash price augmented by the interest rate over the term considered and diminished by the borrowing cost of the metal. The futures price of gold was thus always above the cash price – sometimes by a large proportion if the interest rates were high. Financing might also take the form of a loan of gold to the mining company. The gold borrowed was then sold for cash in dollars, releasing the resources necessary. The gold was reimbursed after the mine was in production for a certain number of years.

Producers used derivative instruments of varying complexity. The simplest consisted of sales of call options with determined terms and prices, and put options designed to ensure protection against an eventual drop in the price of gold. There was also the zero-price "collar" instrument: a purchase of puts matched with a simultaneous sale of calls (the latter having a higher price than the former). The most complicated combined purchases, sales, different terms, threshold effects, and so on; these were termed "exotic." Sometimes, the complexity was increased by including currency transactions, particularly for companies whose costs were established in currencies different from the currency in which production would be sold, the U.S. dollar. These products, traded on over-the-counter markets and designed as a function of a company's needs or according to the unbounded imagination of intermediaries, were used massively in 1997 and 1999.

The "accident" occurred in September 1999 when the European central banks agreed to limit their sales, with the limitation extending also to gold loans. The result was a strong increase in the interest rate on these loans, which for a time rose above the London Inter-Bank Offered Rate (LIBOR) – that is, futures sales were taking place at a price below the cash price. "Exotic" derivative products thus became very risky.

For some companies, which had given in to the temptation of speculation, the situation became even more difficult: unable to deliver the quantities that they had sold, which were more than they had produced, they suffered considerable losses. Among the consequences of this situation were simplification of the derivative products used by mining companies and a subsequent rise in gold prices. Due to their complexity derivative products were difficult to analyze, even for experienced experts.

Today, the trend is completely reversed. In an inflationary environment, the mining companies are eliminating their systems of protection against a drop in gold prices by buying back their futures contracts or delivering on existing contracts ("dehedging"); in a direct or indirect way, this is contributing to the demand for about 300 tons per year (average for the last five years). The extended pick-up in metal prices has not yet been entirely exploited, to the detriment of operating margins and discontent of shareholders, while obtaining loans is becoming easier.

The central banks are still the first interveners on the gold market. They ensure its liquidity by lending metal, in general through specialists called bullion dealers. These loans are indispensable to the running of some operations or the development of derivative products and are made at rates that fluctuate according to supply and demand. For certain central banks, the ability to make loans depends on their prudential ratio. Thus, some central banks do not make loans, while some act more boldly, increasing the risk of offsets.

These loans also provide an opportunity to ensure low profitability on the stock held. The

Trading floor of the New York Stock Exchange
Photograph: © Erik Freeland/Corbis

central banks (including the IMF) still hold about 30,000 metric tons of metal in their exchange and gold reserves out of a global store, discountable or not, of about 150,000 metric tons. The distribution of the metal is unbalanced.

At the end of 2006, the central banks that were signatories to the agreement to limit sales held 42% of the world stock: the United States held 2%; the IMF, 11%; China, 2%; and the rest of the world, 18% (rounded figures). The European central banks supply the market with about 500 metric tons. Transfers by central banks are increasing the surplus already observed in the primary market (mining production + recycling – jewellery + industry); the combined surplus reached a total of 1,100 tons in 2007.

Jewellers and industries have little effect on the market, as their operations are simple and devoted essentially to management of their stock. The combination of investment and speculation has thus for several years been the engine of the gold market. Investment in gold in its traditional form (bars, ingots, coins, commemorative coins) is still important in the Middle East and Far East, but it is running into a number of limitations (hoarding, insurance, manipulation) and the regulatory ban on certain organizations (pension funds, etc.)

purchasing physical metal. Investment in "paper gold" has thus developed. Specialized funds have been created ("exchange traded funds" [ETFs]); listed on stock exchanges and having the status of securities, they exist solely to manage the volume of a gold store as a function of the supply of and demand for its stock. The first of these funds was launched in Australia in 2003. At the end of 2006, there were seven quoted on nine markets. Their success made them worth outstandings of 648 tons, and they had grown further by the end of May 2007. It is foreseeable that more ETFs will be created, given private projects concerning India, a great importer of gold, and Japan. This success should be contagious and other funds will be launched. Neighbouring products already exist (Canadian funds, various certificates, etc.). The success of index funds (ETFs) also owes much to their relative simplicity of operation (good duplication of metal prices, modest operating costs). The outstandings of these index funds are fairly sensitive to fluctuations in the price of gold, to the point that stakeholders are mostly medium- and long-term investors. Shorter-term speculation is exercised on traditional markets or, for some structured products, on the over-the-counter or unlisted securities markets. The main markets for listed instruments linked to gold are the COMEX and the Chicago Board of Trade in the United States and the TOCOM in Japan, where contracts with various maturity dates and options are traded. These markets, however, are not the only ones. Other, less sophisticated ones have been created in Shanghai (Shanghai Gold Exchange), India (Multi Commodity Exchange, National Commodity, and Derivatives Exchange Limited) and Dubai. The American markets are transparent, and the speculative positions are known (combined non-commercial and non-reportable positions). These positions are an excellent indicator of the general level of speculation on the gold market.

There are many possible combinations to be made from these standardized products. One of the most notable is exploitation of the correlation between fluctuations in the dollar and in gold. There is a "mirror effect" with a correlation that may reach 0.9 to 0.95 except in extreme circumstances. Gold may thus be used as protection against fluctuations in American currency without being the only instrument available, and with the understanding that symmetry does not mean equality of amplitude.

Activity on the over-the-counter and unlisted securities markets is expanding, in some cases to the detriment of the traditional futures markets. A variety of products are traded – various derivatives, including, mostly, products structured to demand, as a function of the operators' investment objectives; some mining producers are active in these markets. They are flexible and relatively low cost; they also have a lack of transparency, which, paradoxically, may be in the interest of some institutional investors. There are all sorts of original products, but also products with fixed rates linked to gold, as well as structured instruments with baskets of raw materials, in which gold is represented according to predefined criteria. These have an important influence on the content of metal, especially when the general trend of the market in raw materials is freewheeling (the boom of May 2006, followed by a sudden drop). Thus, the correlation between these markets and that of gold has become clear, which corresponds to an evolution characteristic of the metals market.

Today, gold markets are developed and sophisticated, which influences the volume of transactions. It is estimated that the volume traded, in all forms of transactions (physical, paper gold, derivatives, etc.) represents the equivalent of 10 million to 20 million ounces per day, or 300 to 600 metric tons (the equivalent of US$6 billion to US$13 billion), compared to the density of the primary gold market (mining production + recycling) of about 3,500 metric tons per year. The gold market is therefore particularly active, but it is narrow compared to transactions on the foreign exchange markets, estimated at a daily average of US$2 trillion (equivalent).

Conclusion and Prospects

Gold has retained an important financial role: that of store of value. Humankind needs a last-resort financial store, accepted by everyone everywhere, guaranteed by its physical characteristics. This may be summarized by the saying "One can print bank notes but not gold ingots." Gold thus remains a refuge value here and there, but it has acquired a status of prudential diversification store in modern economies at the same time as its use has evolved as a function of its technological characteristics, thus leading to an active and sophisticated market. This market remains narrow, which impedes the acquisition or disposition of physical metal by important stakeholders. The prospects for the metal seem to be ensured over the long term, even if one admits that its monetary role is in the past. Its characteristic as a hedge against inflation, monetary devaluation, and other events remains undeniable and make it attractive. Its main handicap, however, is an absence of returns.

Bibliography

Bernstein, Peter L. 2007. *Le pouvoir de l'or, histoire d'une obsession.* Édition Mazarine.

Eluère, Christiane. 1990. *Les secrets de l'or antique.* "Aspects de l'art" Coll. Paris: Bibliothèque des arts.

Documentation UBS (Swiss organization for financial management for individuals and enterprises)

Documentation Global Gestion

Gold Surveys: GFMS Limited

The Royal Canadian Mint

IAN E. BENNETT
President and CEO, Royal Canadian Mint

GOLD, RARE, LEGENDARY, AND precious, has fascinated humanity since the dawn of recorded history. It continues to captivate us at the Royal Canadian Mint, where refining the purest gold and crafting the finest gold bullion and collector coins in the world matches our passion for producing high-quality circulation coins for Canada and over 60 countries around the world.

The lure of the noble metal is easily understood. It is a universal expression of value and never tarnishes in its pure state. It is the most malleable of metals and the only one that can be made thin enough to become transparent. A single one of our new 99.999% pure gold bullion one-ounce coins could be stretched into a thin, continuous gold thread 80 kilometres long. This amazing element's properties have made it a coveted substance for applications in art, industry, and science.

It has been said that gold is the king of metals and the metal of kings. Since 1908, its presence in the fabric of our history has crowned the Royal Canadian Mint with many landmark successes in refining and manufacturing gold. We are proud to display several of our accomplishments in the *Gold in the Americas* exhibition.

Our contribution adds one more precious thread to the rich tapestry of artefacts brought to Quebec City by the Musée de la civilisation, which tells the captivating story of gold in North and South America, from its influence on our early cultures to the prominence that it commands in today's business world.

Maple Leaf bullion coin weighing 100 kg
Dated 2007, in 99.999% pure gold, coin, with a face value of $1 million, is surrounded by pure gold bars, wafers, and coins. This coin was certified by Guinness World Records as the largest gold coin in the world.

Created by Stan Witten
Diam: 51 cm Th: 3 cm

◄ *Raven Bringing Light to the World*
Gold mask by Robert Davidson.
This mask was the inspiration for the 22-karat $200 gold coin struck by the Royal Canadian Mint in 1997.

Diam.: 30 cm

The morbid delirium of mines manifested its first symptoms. it is not easy to define this sort of possession that works subtly on us yet has nothing to do with the very real desire to discover extraordinary riches. This desire is not the main motive; there's something deeper to it, more indefinable. Something linked to the gold, of course, but simply because we extract it from Earth, because Earth holds on to it jealously, releasing it only after a hard-fought battle in which our very skin is at stake. It's as if we wanted to hold in our hands, just once, a minuscule and unfortunate fragment of eternity. This has nothing to do with the power exercised by a drug, nor with the fascination of the game.

Alvaro Mutis
Mutis, Alvaro. 1992. Écoute-moi, Amirbar. Paris: Grasset.

CHAPTER **9**

GOLD: MYTH AND REALITY

Gold and a Brief History of Eternity

Jacques-M. Chevalier, Ph.D. Anthropology, Carleton University, Ottawa,
and Zélie Larose-Chevalier, anthropologist and museologist

Gold and a Brief History of Eternity

JACQUES-M. CHEVALIER
Anthropologist, Carleton University, Ottawa

ZÉLIE LAROSE-CHEVALIER
Anthropologist and museologist

Translated by Joan Irving

GOLD, LIKE MONEY, DOES not make you happy. Dreaming the reverse, Croesus, king of Lydia, tried one day to impress an Athenian by the name of Solon with his gold. The sole response he got was, given the jealousy of the gods, the only thing that had any bearing on happiness was chance. Midas, a Phrygian king who dreamed of being as rich as Croesus, learned the same lesson thanks to the benevolence of the god Bacchus. According to the legend, Bacchus told Midas that he might have anything he wished for. The delighted Midas asked that anything he touch turn to gold. But the granting of his wish became a nightmare when his meal and his daughter, the first things he touched, turned into gold! Lydians or Phrygians, kings or peasants, alchemists or beggars – all who believe that gold is the key to eternal happiness are reminded of the imperatives of chance and life as we know them.

And with reason, for it's not gold that writes history, but rather the history of the world and the commerce of desire that have shaped gold. The history of gold and of the desire it embodies take us back to three stories about eternity that have contributed to the ongoing globalization of the world, that is, to the formation of empires and to their economic and political transactions on a global scale. First there was the ancient world that created the bimetallic monetary system and surrounded it with the aura of the sacred – the flesh of gods and the golden calves of Antiquity, golden bridges between the eternal and all who believed in it. Then came the gold ingots and gold standard of the New World, measures of wealth that kept a certain distance from the sacred and engraved their value in the permanence of nature,

◄ *Dawn of the Gods*

Thésée, Québec, 1990
Photograph: Idra Labrie, Perspective/Musée de la civilisation

stable commerce, and secure markets. The last push toward globalization, managed in the American way and by far the most unrestrained, is expressed through floating gold rates, a business model for gold that precariously straddles postmodernism and the desire for the ephemeral and its eternal present.

Every formula for eternity forged in gold has dehumanized that which belongs to the history of the world and to the empire of desire. In this sense, humanity has yet to construct a form of "globalization" grounded in worldly history. This anthropological and ever-evolving project will come to pass when humanity accepts its own history and its liberty in fullness and wisdom, without the numerous crimes committed in its name.

The Ancient World:
Flesh of the Gods and of Eternity

The use of gold and silver as units of tally and repositories of value goes back thousands of years. Once the sole privilege of kings and pharaohs, these two metals have been used in business deals since ancient Egypt and, even before that, in Akkad (central Mesopotamia). Gold assumed its true monetary power more than 2,500 years ago, on the day that Lydia created its state monopoly over the production of currency and cut the first gold coins, which could be converted into smaller silver coins. Because gold was less practical for everyday transactions, silver remained the most frequently used metal in monetary exchanges right up until the twentieth century. The bimetallic monetary system invented in Lydian territory was thus the beginning of the international history of measurements of wealth.

Gold played a central role in the beginning of globalization. Although different from the accelerated globalization we are now experiencing, the push toward globalization that traversed Antiquity enabled

Pectoral, Veraguas, Panamá
Musée des beaux-arts de Montréal, Inv. 1960.Ac.1
H: 1.8 cm Diam.: 17.2 cm
Photograph: Christine Guest/Musée des beaux-arts de Montréal

the development of empires, markets, and networks of economic interdependence that encompassed the known world. Gold is thus the key to history on a large scale and over a long period of time. It becomes all the more encompassing in that it is synonymous with military campaigns and territorial expansions, through the payments to mercenaries and armies bent on conquering the world fallen prey to pillage and the horrors of war.

The explanation as to why gold is so useful as a measure of wealth lies in its physical properties. The metal is rare, dense, and extremely long lasting, while at the same time identifiable, malleable, and easy to melt. This down-to-earth explanation, as comprehensive as it may be, is nonetheless incomplete. The meaning of gold lies also in the realm of desire and the imaginary. Although it is a metal extracted from the earth, gold has a brilliance, a purity, and a durability associated with the divine and the privilege of celestial beings. Numerous were the emperors, kings, and princes who gilded their residences as well as their gods! All of the temples, statues, and monuments dedicated to the glory of the gods; all of the chariots, thrones, sceptres, mirrors, jewellery, and illuminations reaching back to high antiquity – what do they have in common? Could they have all put up at the sign of the divine? Didn't Yahweh announce to Moses that the Egyptians would be fleeced of their gold and the flesh of their gods so that the sons and daughters of the chosen people could don it and devote themselves to the building of Solomon's temple? Were not gold coins used to print the image of Christ under the pious reign of Justinian II (685–95, 705–11) and, later on, the biblical word under the papal reign of

Innocent IV (1243–54)? Did gold not also serve to transmit the word of Allah in the Islamic world of the late seventh century, at the official birth of the dinar?

Several ancient civilizations used gold as a measure of the divine. The Incas, the Aztecs, and the Mayas did so long before they were conquered. In their case, as with many others, the tendency to surround the divine in shimmering gold was linked to the cult of the Sun, the celestial corps that, when the world became the world, established the temporal order as we know it. The word "gold" comes from the Latin *aurum*, meaning also the dawn. And of course a brilliant yellow sun-god who lived at the centre of the universe was the inspiration for those who worshipped the golden calf and whom Moses denounced after his meeting with Yahweh on the summit of Mount Sinai. It is to this same celestial body that we owe the bull and lion engraved on Lydian coins. These golden animals are astrological signs symbolizing the resurrection of the Sun at the spring equinox (Taurus, the bull) on the one hand, and its highest point during the summer solstice (Lion) on the other hand. Given the planetary and zodiacal antecedents of gold worship, we shouldn't be surprised at the importance of the duodecimal (12) numbering system, handed down from the Chaldeans, that has governed numerous monetary systems for thousands of years: the solidus worth 12 dinars, the troy ounce composed of 24 karats of pure gold, and, closer to home, the English shilling worth 12 pence.

Thus, from earliest antiquity, gold served as the standard for life that was eternal, infinitely malleable and durable, timeless and pure. Gold is the hoarding of time conjugated to infinity. In the beginning, globalization was set off through what could be called the unworldliness of gold, that is, its constant flight toward another world. Of all the paradoxes invented by humans, eternity conceived as the basis of the law of temporal and material exchange is by far the most surprising. The mirror effect of this paradox is just as surprising. Nothing can be deemed immutable or eternal without a carnal world that keeps up its allegiance with the sacred and ensures the trade in faith, investments, and speculations in the hereafter. This surreal commerce is a response to the jealousy

Funerary mask, Nasca, Peru
Even covered with gold, the princes of this world and their gods must someday die. The function of the mask is to transform the one who wears it during his voyage to the hereafter.

Musée des beaux-arts de Montréal, Inv. 1972.Ad.1
H: 13 cm W: 20 cm
Photograph: Christine Guest/Musée des beaux-arts de Montréal

and folly of the gods, who cannot get along without the loyalty and sacrifice of human beings, even unto their death. After all, the immortality of some depends on the mortality of others. The quest for infinity by the all-powerful hinges on the desire for eternal life among those who know they will never be kings or God's heirs. The infinite is convincing for those who possess it only if those who are excluded believe and desire it, that is, envy the gods and their protégés for their lives and their gilded days. Material commerce and faith in the beyond are just two sides of the same ancient coin. Faith and gold as cash both serve as guarantees in the commerce between gods, kings, and humans.

The fate of this world is linked to the gold of the gods. Because of this worship of gold as absolute wealth, the whole business of vows and offerings acquired its overwhelming materiality and turned into supply and demand. But the desire for the absolute is simply the absolute reign of desire. A fluctuating reign, if ever there was one... Like any kingdom, the empire of desire has its own history, and also countless sorrows and sufferings! Even when they are covered in gold, the princes of this world and their gods eventually die out. The only thing that remains is the order of desire, guided by the signs and the

desiderata (Lat. *desiderare, de-* + *sideris,* "star") of what we await from the Sun and the stars. Isn't this the sole precious item that the dead and the gods envy us for, our desire for another world that persists and simply lives on upon the death of kings and gods?

The New World:
The Gold Ingot and Perenniality

Gold is at the cusp of the temporal and the eternal, two parts of a single clasp inevitably corroded by time. By Creosus' time the gap between gold and the eternal was growing and gold was starting to democratize: it became coinage and, therefore, accessible to common mortals. Next came the despoilment of the gods, as seen in the stripping of the fabulous golden robe of Athena, goddess of wisdom and the Parthenon, to finance the cost of a military campaign, an act of despoilment that has resonated across the centuries. An awful lot of gold was needed to finance the Crusades and the military campaigns of the Middle Ages. The so-called heathen temples and their golden treasures were pillaged during the reign of Constantine I, who converted to Christianity (306–37), and also Henry VIII (1509–47), in his fight against the Catholic Church. Iconoclasm, adopted into law by Leo III during his war-faring reign, was equally destructive to sacred objects and lasted more than one hundred years (730–843).

Of all the pillages of gold that history has ever known, that of the New World remains unequalled. Before its discovery of America, the ancient world had created a whole story of the Beyond woven in gold, to such an extent that everyone believed the world was nearing its end and about to melt away in the eternal. Flouting innumerable apocalyptic prophecies, the world entered a period of triumphant expansion and survived, thus sounding the knell of eternity and its sacred gold. The story of the fall of eternal gold starts with the death of the Inca emperor Atahualpa at the hands of the intrepid and villainous Francisco Pizarro – a human tragedy, certainly, but also a divine one, in which the son of the Sun is assassinated and his votive objects melted into vulgar gold ingots. The discourse on the holy war hawked by Pizarro and his missionary associates left a lingering odour to be sure,

but his brutal conquering act and the Renaissance (1420–1630) of the world that accompanied it brought an end to the short history of eternity. The followers of God gave way to a stampede of gold-thirsty explorers like the great explorers Henry the Navigator, Christopher Columbus, and Ferdinand Magellan. Their world now ancient, the pursuit of eternal life made way for royal and princely power that, although a divine right, adorned itself with the thrones, crowns, sceptres, palaces, jewels, and fabrics that were no longer the purview of the gods.

Gold, which was sorely lacking in medieval Europe, resumed its role as the engine of economic and territorial conquests thanks to the gold resources of the New World. The value of the precious metal for ostentation was also quickly restored, but this time at some distance from the divine. The kings of Europe appropriated the gold of the New World for themselves, making it both the symbol of their absolute power and the underwriter of their riches and immortal memory. What followed was a display of gold not seen since the lavish reign of Egypt's Hatsheput (1470–58 BCE). The Field of the Cloth of Gold, in Flanders (1520), is the most famous example; it, all on its own, signified the end of the Middle Ages. The meeting site is renowned as a demonstration of the power and mutual rivalry of François I, king of France, and Henri VIII, king of England. The two kings and their courts rivalled each other in shows of wealth and finery there, each displaying hundreds of tents covered with fabrics embroidered with gold thread. The French royal tent was draped in gold and England's crystal palace, no less sumptuous, was

Henry VIII and Francis I Meet on the Field of the Cloth of Gold Near Calais
Both monarchs' tent canvases were woven with gold threads and rivalled each other for opulence.

Friedrich Bouterwek, after Hans Holbein the Elder, 1520
Chateau Versailles and Chateau Trianon, Versailles, France
Photograph: Erich Lessing/Art Resource, New York

decorated with gold frieze and displayed inside a tapestry covered with golden lilies.

The imaginary that launched both the New World and the Renaissance no longer rested on the faith of a people in search of eternal life, but henceforth on the two-fold commerce of desire: on the one hand, a world of plenty and immortality symbolized by the royal splendour and, on the other hand, the safety and order that the sovereign promised his subjects as a way to escape the misery, war, and torture that haunted the Middle Ages.

The gold rush in the Americas, sponsored by the kings of Europe, was the epicentre of the earthquake that the coming of the New World represented. The race to power in the New World was the most globalizing struggle for power ever seen. Gold as wealth triumphed over its age-old adversary, sacred gold. As of the eighteenth century, the emerging "modern" world would retain nothing but the reassuring aspect of the timeless, that of marketable desire, prepaid and guaranteed transactions, the order and stability of forever rising markets. The value of gold is absolute, in the modern sense, because it comes from the permanence of nature, thus ensuring the perenniality of material happiness, on which time has no bearing. Gold is now detached from worldly history not because of its sacred function but because of its natural value, which transcends time.

As with eternity, the story of the timeless turning into material abundance and natural security is an expression of desire and lasts but a moment. The gold of the New World brought days of glory as well as descents into hell. Among the latter were the numerous gold shortages that led to economic and territorial crises which still haunt the modern world. These shortages occurred because of the inevitable exhaustion of the gold mines, and because of individual and public hoarding that removed gold from the market in reaction to the insecurity created by wars, famines, epidemics, and wide-scale depressions and economic crises.

The scarcity and shortages of gold in modern times led to the passing of all sorts of laws and remedial measures, both ingenious and awkward, aimed at ending the insecurity and arbitrariness of markets fallen victim to inflation and depression. One such measure was the gold standard. It fixed the value of every national currency in relationship to gold, which served as the reserve asset for the central banks and also the international currency for regulating trade. A system put in place in England at the beginning of the nineteenth century, it was globalized in under a hundred years, ending with one stroke the long monetary rivalry between gold and silver.

In theory, the gold standard represents the foolproofing of a market economy determined to protect itself from both the uncontrolled expansion of credit and the inflation that results from the devaluation of the national currency. When there is rapid economic expansion and inflation, the banks can protect their gold reserves under pressure by increasing interest rates to attract new capital. This increases unemployment, but gradually raises the gold reserves and

improves the exchange rate and therefore the value in gold of the national currency. The State has no need to intervene; it can count on the value of its gold, the natural and absolute source of all wealth, remaining stable, and on the invisible hand of the laws of the market as described by Adam Smith (1723–90). Thus does the State ensure the stability and growth of its markets, and also that demand will never exceed the absolute wealth at its disposal.

It took the Depression of the 1930s to prove what John Maynard Keynes and President Roosevelt both considered to be the absurdity of the gold standard. They argued against the widely held and thoroughly mad belief according to which gold, by its very nature, makes things easy for all nations and solves all problems, as long as it is wisely stockpiled and securely stored. This idea underlines the difficult task of a world that is dragging its feet in writing its own history. It stems from the make-believe world of gold as absolute and timeless that has held sway for several millennia. And yet, absolute gold and its modern heir, the gold standard, are not eternal. Recent history is sounding their knell, at the dawn of an ever more globalizing era.

The Postmodern World: The Golden Opportunity and the Eternal Present

The postmodern period lends credibility to a new form of timelessness that is once again holding back a worldly globalization understood as the conscience of humanity fully present to its own history and times. It highlights the growth of the market and its powerful engine, unlimited demand, which any measure of value or metallic standard would be hard pressed to accommodate. Metallic currency has given way to paper money, to cheques and bank accounts and electronic banking, fiduciary rivals that emerged during the Renaissance and made it possible for trade to expand well beyond what the gold reserves could support.

The globalization of today's borderless markets is based on aggressive marketing to whip up instantaneous and fleeting consumption. Infinite happiness zapped on the whim of the eternal present now reigns. The desire for eternity, immortality, and security has given way to golden opportunities advertised through invasive publicity. Humanity lives from day to day, dragging its feet on the project of realizing a world that makes its own decisions and writes

Sharp Edge, interior view

Daniel Brush, 1995–96
Pure gold and steel
Private collection
Photograph: David Behl

its own history. Everything floats in the moment, including gold.

One might say that the postmodern period started in 1944 with the Bretton Woods agreements by which the gold standard was replaced by the Gold Exchange Standard that made the U.S. dollar the only currency convertible to gold and fixed the rate of exchange of other currencies to the dollar. These agreements remained in force until the member governments of the "Gold Pool" decided, in 1968, to stop guaranteeing a fixed price for gold and to let it fluctuate in the marketplace, thereby dethroning the U.S. dollar. In 1971 the Americans retaliated, taking the whole world by surprise. To control inflation, reduce imports, and attract foreign investment, the United States, which was short of gold, unilaterally renounced the convertibility of the dollar. A new system of floating exchange rates was launched in the 1970s; the value of national currencies would henceforth float, determined by the vagaries of supply and demand. The price of gold would be subject to the same fate.

Over the past two decades, the value of gold has fluctuated in relation to the inventory of mineral reserves, the costs and volumes of production, the evolution of the gold reserves of the central banks, the needs of industry, and the needs of goldsmiths. Gold of course still serves as a hedge in times of crisis. Despite its highs and lows, the price of gold has in general been falling; between 1980 and 1999, it declined 60 percent. But the fact that gold has been selling at a lower price has led to an increase in demand, which doubled in the 1990s. At the same time, the jewellery industry enjoyed unprecedented growth.

The dollar resumes its dominance, and gold becomes, for the first time in its history, glittery merchandise aimed at the consumer who is, apparently, king. Despite this, gold is losing its purity and its panache, even in ceremonial situations. Gold is now accessible, varied, popular. Since the 1980s, gold jewellery, watches, pens, eyeglass frames, and bathroom accessories have been available for purchase over the counter in big box stores. The quality of the gold varies widely and so do the prices, styles, and colours. Unconcerned whether they are wearing a metal in its

Gold necklace

Designed by Paloma Picasso
Photograph: © Tiffany & Co.
Archives 2008

pure state, today's consumers want long-lasting alloys in a colour that matches their complexion or their clothes. White gold (an alloy with silver) and pink gold (an alloy with copper and silver) are the most popular of the many available shades, including red gold (copper), green gold (nickel), grey gold (iron), and violet gold (aluminum).

In the twenty-first century, gold has taken on new forms and become so popular that it is déclassé. Gold has lost ground to diamonds and platinum. Wedding rings, now fashioned of white gold or platinum more often than yellow gold, are often adorned with a diamond that attracts more attention than the ring itself. The platinum anniversary is ten years more "valuable" than the diamond anniversary, and twenty more than the golden anniversary. On the Canadian hit parade, an album is said to have gone diamond when one million are sold; this represents sales ten times greater than for an album that has gone platinum and twenty times greater than for one that has gone gold.

The last straw is that gold is becoming useful. Thanks to its malleability, ductility, reflexivity, resistance to corrosion, and exceptional electrical and thermal conductibility, gold is now prized in numerous industries. Without gold, practical objects such as the telephone, computer, calculator, washing machine, television, and airplane would not be what they are.

The more useful gold is, the more it is regarded as harmful and is subject to criticism. The list of crimes committed in the name of gold grows longer and longer. Gold mines devastate ecosystems as well as community traditions and lifestyles. Gold's image is becoming tarnished, a situation countered by the marketing efforts of the World Gold Council, which dubs itself the "Global Advocate for Gold."

Gold has lost its shine and become a shadow of its former self. No longer pure and useless, it now finds itself in several alloys and very useful, although not as sophisticated as some other glitzy and industrial metals. Because its value is no longer guaranteed, gold floats. It's a true reflection of its time, a postmodernism floating in the limbo of a world transformed by globalization and governed by the kings of black gold, finance, and marketing. The gold of today occupies a poorly defined and little explored niche of history, a liminal era that brings together all of the timeless illusions and fears of our past. Postmodernism contains the universal legacy of the old hopes and anguishes of the hereafter, the permanence of nature, and life lived henceforth in the immediate present. The longing for eternity has gone by, but humanity is once again caught up in the jealousy of the gods and the zealousness of a Free World taking on the Axis of Evil or Islam, declar itself, as in times past, at war against armies of crusaders. Starting with the Renaissance, eternity gave way to the permanence of Nature – a natural world that stood as the source of all riches and remained, until our time, the great rallying point for life. But the world in its so-called natural state is threatened by industry, now more polluting than ever, and money-grubbing biotechnology, prepared to modify the genetic code as it sees fit. We have now reached a stage where the order of desire and the imaginary fluctuates daily, governed as it is by signs of golden opportunity and short-term promise. But security and the growth of the markets are once again threatened by the spectre of war, hardship, epidemics, and planetary catastrophes, real or potential – by many long-term dangers that reduce postmodernism to this very brief moment we will nevertheless live as the longest in our history. As Woody Allen reminds us, "Eternity is really long. Especially near the end."

Could it be that the liminal era of floating gold is marking the end of world history, a limbo world that brings together and globalizes the impossible dreams and nightmares from all times? To think of it, the reverse is closer to the truth: after several millennia of eternity, the history of the "world in this world" is finally beginning. Postmodernism and its eternal present are nothing but a jolt among all the timeless eras following one upon the other and never ending, the last breath of a globalization that has been working away for millennia to negate the wordly fate of humans and their commerce of desire.

Floating gold invites us to end the globalization that is being erected to the great detriment of this

world. It proposes a final day that echoes the first day when Creosus and Midas set out in search of absolute and otherworldly happiness. The grand narratives are dying, including the story of gold. Still, gold should not be erased from our memory. It is already worthy of being displayed in a museum, in illuminated manuscripts that remind us of the hard lessons of eternity rendered a thing of the past. The world might gain something from this because, for the first time in history, humanity is called upon to reflect upon itself, without the complacency of the mirror effect. Humanity is urged to unravel its own project as it becomes, in the words of Jean-Jacques Rousseau, "the wise power that governs the world." This is not a power that gives assurances of the absolute or guarantees of a sacred or natural order that transcends history. Neither is it a power that can be handed over to a trust or a single authority that captures and monopolizes the power of trade and threats thereof. For this power to become truly wise, the fiduciary principle, from the Latin *fiducia*, "trust" that is the basis of all human exchange, must take flight and flow to all of humanity, which alone is able to create its own *values* and give new meaning to history.

Gold ingots

Photograph: Joe Drivas/Getty Images

EPILOGUE

GOLDEN TREASURE

The Production of Fake Pre-Hispanic Metal Objects
PALOMA CARCEDO DE MUFARECH, professor of history and art, Pontificia Universidad Católica, Lima, Peru

The Biggest Treasure Chest in the World
HÉLÈNE DIONNE, folklorist and museologist, Head of Publications, Musée de la civilisation

Gold in the Americas: A Voyage Beyond Boundaries
HÉLÈNE DANEAU, historian and museologist, Project Manager for the exhibition *Gold in the Americas*

The Production of Fake Pre-Hispanic Metal Objects

PALOMA CARCEDO DE MUFARECH
Professor, Art History, Pontificia Universidad Católica, Lima, Peru

Translated by Joan Irving, from the French translation by Louis Jolicœur

ARCHAEOLOGISTS AND HISTORIANS WHO study the pre-Columbian art of Peru have become accustomed to the sad reality of tomb robbing, destroyed heritage, forgery, the buying and selling of objects, and other adversities characteristic of this field. If all of this is going on, it's in part because Peru is heir to a rich pre-Hispanic material heritage and has, through the centuries, preserved something that numerous other countries have lost, namely, a rich and varied intangible heritage, transmitted from one generation to the next by artisans.

Numerous studies on metalworking in Andean communities have shown that today's artisans use techniques and implements very similar to those used more than one thousand years ago by the cultures that flourished in the central Andes. A similar phenomenon is found with respect to pre-Hispanic ceramics and textiles, as well as paintings from the colonial period. Unfortunately, the experts who bear the huge responsibility of conserving this

◄ Ceremonial knife (*Tumi*)

Sicán (Lambayeque), Batán Grande, Peru, pre-Columbian, ninth–eleventh centuries CE

Hammered gold, silver, and turquoise

Metropolitan Museum of Art, New York, Jan Mitchell and Sons Collection, gift of Jan Mitchell, 1991, Inv.: 1991.419.58

H: 33 cm

Photograph: ©Justin Kerr/The Metropolitan Museum of Art

intangible culture encounter unscrupulous merchants who offer poorly advised or uninformed collectors items that can be called nothing more than "modern handicrafts."

The problem of the forging and production of copies of Peruvian archaeological metal objects is not new, although it has grown in recent decades. It began in 1937, the year in which *El Comercio* and other newspapers and specialized periodicals published reports on the fabulous discoveries made by tomb robbers: a *tumi*, masks, vases, and other objects from Batán Grande, on Peru's northern coast.[1] The news of the discovery of the region's "hidden treasures" spread like wildfire, and the Sicán mask and *tumi* went on to become the preeminent symbols of pre-Hispanic goldsmithing. Then, on June 15, 1959, the Spanish edition of *Life* magazine published a photograph of an extraordinary Sicán mask – mistaken at the time for a Chimú mask – on display at the D'Arcy art gallery in New York.[2] When this photo first appeared, there were many questions about the object: Was it a fake or authentic, and how had it been taken out of Peru? An American by the name of James W. Moseley was even accused of having secreted it away.[3] In the wake of the scandal that erupted over the mask at the D'Arcy gallery, *El Comercio* and other local newspapers began publishing articles on the seriousness of the situation.[4] Similar masks started showing up in other parts of Peru, in both

1. "Los trabajos arqueológicos en el departamento de Lambayeque," *El Comercio*, Lima; "El Oro de Batán Grande," *Revista del Museo Nacional* 6, Lima, 1937, pp. 144–68.
2. *El Comercio*: "Desmedro de nuestro patrimonio arqueológico," 15 June 1959, and "El Perú exigirá la devolución de la mascara de oro," 17 June1959.
3. *Peruvian Times*, 6 May 1960, p. 3.
4. *El Comercio*: "No aclaran como apareció en Galería de los EEUU valiosa máscara incaica," 15 June 1959, and "El misterio de la máscara de oro," 20 June 1959. *La Prensa*: "Se cree son muchas las máscaras de oro que han salido del Perú en forma ilegal," and "El Perú exigirá la devolución de la máscara de oro," 17 June 1959.

The sacrificed youth and the Gold Chest in the Huaca Loro East Tomb

Burial Chamber of the East Tomb at the north base of the Huaca Loro Temple mound at the site of Sicán,
mid-Le Leche Valley, north coast of Peru. Photograph: Izumi Shimada

national museums and private collections, such as that of Hugo Cohen,[5] which was later purchased by the Wiese Bank and is today on exhibit at the Museo del Banco Central de Reserva in Lima.

The Museo del Banco Central de Reserva collection is particularly problematic. It contains a large number of fakes, despite the fact that a short time before his death the collector Raúl Apesteguía[6] made a selection aimed at eliminating the fakes and ensuring that only the "authentic" objects would be put on display.[7] By late 1959, Sicán masks had grown so popular that articles and cartoons were appearing in local newspapers on the large number of forgeries of them throughout Peru.[8]

From that moment on, forgeries of objects from Sicán culture, namely, masks, vases, *tumis*, and items of finery similar to those pictured in the catalogues of international exhibitions, began showing up for sale, particularly in Peru and the United States. The proliferation of the forgeries was so great during the 1960s that American and European collectors began writing to various experts and sending them photographs of items available for sale. One gallery specializing in the sale of authentic pre-Columbian metal works from Peru, in particular those from the Sicán and Chimú cultures, was the André Emmerich Gallery in New York. The high demand for and value of such works only encouraged the forgers!

I worked as an assistant to Junius Bird from 1979 to 1984; during those years, we became alarmed at the information we were receiving

---

5.   *El Comercio*: "Cuatro máscaras de oro mochica posee el coleccionista Hugo Cohen," 22 June 1959.
6.   Raúl Apesteguía was assasinated at his home in Lima on January 26, 1996.
7.   There are still fakes in the collection of the Banco Central de Reserva, as far as I am aware.
8.   *La Prensa*: "Ratifican que aquí se hacen réplicas de máscaras de Oro," 8 November 1959; for a cartoon, see *La Prensa*, 11 August 1959.

Vase, relief decoration portraying
a human head with feline canines,
almond-shaped eyes, and ear pendants.

Sicán (Lambayeque), Peru, ninth–twelfth centuries CE
Museo Arqueológico Rafael Larco Herrera, Inv. ML100759
Photograph: Museo Arqueológico Rafael Larco Herrera, Lima, Peru

about forgeries and we designated four categories of forgery of pre-Hispanic metal objects from Peru. Professor Bird baptized them the "Mr. Chan-Kham collection," "Punch'n Judy," "Pink Crud," and "Bonamichi," the last being the name of a forger killed by a gunman on a street in Lima in 1964. The greatest of the forgers, the Lanao brothers, who specialized in Sicán masks and vases, are unfortunately no longer alive.

Until 1987, the forgers preferred to copy works from the Sicán and Chimú cultures. They based their work on items that were discovered or pillaged from various sites: the Mochica tombs of Huaca Rajada (May 1987) and La Mina (1986–88), the tombs of the Sicán elite at Huaca del Oro in Batán Grande (1991), the Cupisnique tombs at Kuntur Wasi (1995). As representations of these metal objects began appearing in art books and monograms, the forgers started using them as models.

In the 1990s and the early twenty-first century, the government of Peru brought in stricter laws to counter the illegal traffic in archaeological works. Although experts now had more detailed information on Cupisnique, Mochica, and Sicán metalwork than during the 1950s, the fakes continued to proliferate, most being produced to meet the demand in Peru. The new forgeries were, however, much more crudely produced. The increase in demand was also being fuelled by a jump in Internet sales; it was now easier to purchase pre-Columbian objects online.

Although the trade in and forging of metal objects are not new, they have vastly increased over the past few decades, mainly because of the following:

1. Photographs of Andean metal objects are rare and access to them is limited. They appear in difficult-to-acquire specialty publications and in high-quality books to which the general public may not have access. Because knowledge about such items is limited, it is easier to dupe consumers.

2. Few archaeologists and experts on pre-Columbian Andean societies have studied metallurgy, a field that attracts mainly scientists and engineers. Archaeology has traditionally been focused on ceramics, architecture, and textiles, and university programs on pre-Columbian metallurgy are few and far between. The lack of knowledge in this area has meant that famous archaeologists have recognized as authentic some metallic objects of questionable provenance now in private collections. This is one of the main reasons for forgeries, and one of the most delicate aspects of the current debate among experts – the fact that an object is reproduced somewhere or exhibited in a museum is not always an indication that it is authentic.

3. Gaining access to pre-Columbian metal collections, whether public or private, is difficult. And the lack of publications and information on the

Funeral ceremony at Huayna Capac
In Johann Ludwig Gottfried,
Newe Welt vnd americanische Historien.
Frankfurt, 1655
Musée de la civilisation, Bibliothèque
du Séminaire de Québec, SQ025651
Photograph: Idra Labrie
Perspective/Musée de la civilisation

museums accepting gifts or seized objects as donations, without allowing professionals to evaluate the donations and report on the advisability of acquiring them.

5. Most collectors of pre-Columbian metal objects are interested in gold or silver objects and, revealing a fundamentally Western mentality, do not understand that gold and silver often served as the base for the application of other elements, such as feathers or paint – to the point that, to make the gold or silver visible, they sometimes remove all iconographic traces on the object. It should be noted as well that the majority of pre-Columbian metal objects from Peru are made from plates joined one to the other or individual objects joined together mechanically; collectors can thus change the pieces, altering them according to "their taste," to make them look the way they think they do in the publications. So, despite the fact that numerous experts, curators, and collectors are aware that forgeries abound and that buying them is illegal, the market for them is stronger than ever.

6. Last – but not least – is the issue of the publications themselves. This is surely the thorniest for experts. Who will cast doubt on the origin of an object that has appeared in a book or catalogue? These publications convey legitimacy. When readers consult a book or catalogue, they don't question the origin of the objects pictured in them because the author of the text is, in a way,

location of the objects reproduced in books makes identifying and analyzing them very hard. In addition, quality objects – gold or silver – exhibited in public institutions are subject to strict security measures. Experts must submit to thorough and lengthy checks before they are given the permits they need to do their research.

4. Pre-Columbian metals are still metals, and "worth their weight in gold." During the twentieth century, entire collections of pre-Hispanic metal objects were used to pay off debts or secure business transactions, not on the basis of their aesthetic or cultural value, but simply for their material value. Some were acquired by banks, and some ended up in the hands of individuals. In most cases, experts were not consulted on these sales, with the result that public and private collections now contain large numbers of objects of questionable provenance. Then there is the problem of government-run

vouching for them. What readers don't know is that experts often send their articles to publishers who may then print them with photos of their choosing, without consulting the author. Or that an expert may write a text for the catalogue of an individual collection without being familiar with the entire collection, although the addition of his or her name becomes an automatic endorsement of that collection. Should experts be presenting reproductions of archaeological metal objects that are unregistered or of questionable provenance? I think it's our duty to demand that all objects published with our backing be legally registered, and to refuse to have our texts endorse objects that are fake or of questionable provenance.

When an object falls into the hands of a collector, it has usually gone through several steps of modification. The first occurs during the looting, that is, the desecration of the archaeological site. In the case of metal objects, this is generally a burial ground or a tomb. Peru has a long tradition of tomb looting, and some of the lootings are done by *huaqueros*, people who specialize in this type of work. They are extremely well organized and might use posts, shovels, levers, and even bulldozers during the job, destroying everything in their way and saving nothing but the objects they consider to have marketable value. With their booty in hand, the looters again modify the objects in preparation for selling. They first wash them to remove any mud or dirt from the tomb. Then, in the case of a mask for example, they remove any remaining cinnabar, paint, and fabric or feathers still attached to it, so that it looks pleasing to them or a future buyer – like a smooth metal object without any other adornment. If when they open a tomb the looters spy semi-precious stones, metal objects, or bits of metal detached from their mountings, they reconstruct them according to their preferences

or those of a future buyer. If these reconstructed objects are reproduced in a publication without any indication that they were altered from their original form, they may then be copied to produce souvenirs, that is, fakes.

Any expert who has participated in a scientific excavation knows that the metal objects in Peru's pre-Columbian tombs are generally found piled one on the other and heavily damaged by the passage of time, running water, the movement and weight of the earth, etcetera, all of which may affect the patina and cause significant chemical transformation. Necklaces are, for example, usually found in a pile, but the cotton string on which they were threaded has often disintegrated. When the individual pieces are quickly gathered, as they would be during a looting, the string breaks and the exact structure of the necklace is destroyed. The looter then reconstitutes the necklaces according to personal taste before selling the finished product to a collector, who displays the necklace as if this were its original structure. The same sort of thing can happen to pieces of fabric decorated with metallic plates. In such cases, the expert must carefully examine the object to try to discern the exact state of the item at the time of burial, and to try to determine its precise use and symbolic role.

The most illuminating case of Peruvian forged metal objects occurred at the Museo de Oro del Perú y Armas del Mundo, in Lima. In April 2001, as an expert with the Pontificia Universidad Católica in Peru, I was asked to evaluate for the Instituto Nacional de Defensa de la Competencia y Protección de la Propiedad Intelectual (INDECOPI) the museum's works in metal. After examining 20,000 objects, my colleague Gabriela Shöerwel, from the Instituto Nacional de Cultura, and I reported five categories of metallic objects on exhibit in the museum's display cases: originals,[9] replicas,[10]

9. By "original" we mean objects that historically belonged to a pre-Columbian culture of the central Andes and that, as a result, have great historical and heritage value. Each object must be exhibited according to current conservation criteria and accompanied by a description of the culture to which it belongs, the material it is made of, the technique used to produce it, and its function. In addition, it must be catalogued and registered in the archives of the museum holding it as well as at the Instituto Nacional de Cultura (INC), based on the standards adopted by that institution.

10. By "replica" we mean objects produced by an artisan using modern-day materials (gold, silver, or copper) and techniques to make an item of the exact same weight, size, and form as that of the original object, to such a degree that visitors to an institution cannot distinguish the original from the replica. For this reason, every replica must bear in some not immediately visible spot the mark of the maker and the date of manufacture. After the making of said replica, the mould, if one was used, must be destroyed in the presence of a notary, because each replica must be unique. It must be clearly identified as a replica "the original of which is found in X museum or Y collection," when on display, in order to prevent any possibility it could be confused with the original. Replicas of metal objects, in particular, those in gold, are on display in museums that do not have the resources necessary to ensure the security of the premises; in other cases replicas are used to help viewers better understand the overall exhibition. Before making a replica, the necessary permits should of course be obtained from the institution holding the original object. Replicas have no historical or heritage value but might have some value in the art market, depending on their quality.

copies,[11] modern-day reconstitutions made using archaeological materials,[12] and modern-day artisanal items or inventions.[13] As illuminating and useful as the nuances between these five categories may be, in reality, they can be reduced to just two: the originals and the fakes.

I certainly found it disheartening that such an important museum had so many fake objects among the unique and extraordinary works on display there. Nonetheless, the experience of determining which ones were originals and which fake was very instructive. First, we realized that the most-copied original objects had appeared in reproduction in books on Peruvian art. We then discovered that the majority of them had been found during scientific archaeological excavations at the sites of the royal tombs of Sipán (1987) and Kuntur Wasi (1992), as well as at the Sicán archaeological project of Huaca del Oro in Batán Grande (1992–95). The value of some of these objects rose sharply after the discoveries and the ensuring media coverage of the events. It is thus easy to deduce when these objects were acquired by the museums.

In art books (but not scholarly books) only the fronts of objects are customarily reproduced, with no indication as to the size or detailing on the reverse side. The goldsmith who must forge an object based on photographs from one of these books has no choice but to guess the exact size and reverse-side detailing. While even without these details the final object may resemble a pre-Columbian object, the expert knows immediately that the goldsmith never had the original in hand.

During this work we also determined that some studios specialized in certain techniques and materials. We observed distinct groups of figurines as well as hollowed, hammered, and laminated objects. The majority of sculpted and hollowed out figurines were done by the same forger or studio in a style, dating from the 1960s, associated with Bonamichi. Most surprising was the fact that numerous objects clearly displayed the workmanship and markings of their makers, to such an extent that after a few days we were able to sort them by artist and by studio. The objects were identifiable not only on the basis of technique, but also of the material used to make them. It was easy to distinguish those who worked in gold and silver from those who used other alloys. The "Bonamichi," "Lanao," "Mr. Chan-Kham collection," "Punch'n Judy," and "Pink Crud" styles were also clearly discernible.

Today's goldsmiths, like those who worked in pre-Columbian times, are specialized artisans, each of whom works in his or her preferred field: hollowed out, laminated, gilded, or welded objects. Certain families or studios may also specialize in working with gold, silver, or copper. Like any artist, forgers of metal works have a style, a trademark. They are creators whose spirit inhabits each work they conceive. I am sure that the goldsmiths who made these objects never imagined that their work would be shown as original pre-Columbian objects.

The experience of the Museo de Oro del Perú y Armas del Mundo should sound an alarm among other public and private institutions. The problem of the forging and the altering of pre-Columbian metal objects is indeed serious, and must be acknowledged. Every expert in Peru encounters this reality. Wealthy individuals buy objects expecting that they have made a good business deal and often find it hard to accept that they have been duped. As for the artisans, they create objects using ancestral techniques and, in most cases, are not to blame for the way these objects are used by third parties.

---

11. By "copy" we mean objects produced contemporarily and based on an original held in a museum or collection. The material used, the technique, and resulting weight, size, and form of the object are not required to exactly match that of the original object. Copies are frequently made now for sale as souvenirs in museum stores. All copies must be accompanied by a description of where it was made and where the original may be found. Copies produced contemporarily have no historical or heritage value but might have some value in the art market, depending on their quality.

12. We mean here objects made of original archaeological components but that may have been modified to re-create an object that may be similar to a pre-Columbian one. These "re-creations," produced according to individual taste, do not adhere to any scientfic criteria or correspond to any iconographic model, technique, or style from the pre-Columbian period. The components of these objects may have a historic or heritage value but the object as a whole does not.

13. We mean here objects produced contemporarily that are neither originals, copies, nor replicas but simply invented modern-day works. The material used, technique, and weight, size, and form of the object have no connection with original pre-Columbian works. These objects have no historical or heritage value, and represent instead what are known as popular handicrafts.

The Biggest Treasure Chest in the World

HÉLÈNE DIONNE

Folklorist and museologist, editor, Musée de la civilisation, Quebec City

Translated by Joan Irving

Chest
Wood and brass hardware
Late nineteenth century
Musée de la civilisation, Inv.: 87-141
Photograph: Pierre Soulard, Musée de la civilisation

CHILDREN ALL OVER THE world dream of treasure. Finding something hidden in an old chest in a dark corner of the attic or buried a hundred steps from an old oak tree – that's adventure, a glimpse into the past. Chests are objects suffused with symbolism and mystery. The conquistadors sent their sovereign his share of their treasure in a chest. The money to pay the soldiers and mercenaries was also kept in a chest, under lock and key. And the royal treasury was always stored in a chest in the safe of a bank. Pirates, bandits, and revolutionaries, from long ago and from today, all lust after just such a

chest. Finally, in the heart of each of us is a chest concealing a treasure that, when we finally locate it, will make us rich and happy. Wishing that you'll find a treasure chest is like allowing yourself to dream of a life of luxury and unsurpassed wealth. Many adults still dream of becoming instantly rich, as revealed by the quasi-universal infatuation with gambling and lotteries. The myth of El Dorado lives on!

But what is a treasure? To whom does it belong? The contemporary definition of a treasure would be a collection of valuable goods that are usually hidden carefully away. A treasure can also be one or several objects that we really admire, that make us dream, and that have great significance for us – objects that are important enough that we want to keep and protect them, sometimes even guard jealously and hope to pass on to the next generation. The word "treasure" comes from the Latin *thesaurus*, a word still used in English to mean a collection of knowledge on specific subjects. Parsing the word we find the root "au." "Au" is the chemical symbol for gold, which for most people is the most precious metal in the world, the ultimate symbol of wealth.

Government authorities in France, Canada, and several other countries in the Western world at one time defined treasure as "any hidden or buried item that no one can claim as personal property and that is discovered purely by chance." To this they added: "A treasure belongs to the person who finds it if the ground in which it was found is not the property of any individual. If someone can claim ownership of the property where the treasure was found, that person may claim half of the value of the discovery."[1]

Passion, desire, and fascination are all fundamental to any treasure. The very word "treasure" conjures up incredible riches and value. Finding an object that you consider to be a treasure deep in a cave, in your grandmother's attic, or at a flea market is to delve into the emotional. It involves transcending the physical materiality of the object and crossing the bridge between the past and the present. To show this object, allow a friend or family member to discover it, to exhibit it and let it be admired is to share the excitement of the discovery, the joy of possession, and finally to pass on some of the passion for this treasure. By according such value to one or several objects, we become morally responsible for ensuring their safety. But to protect treasure we must withdraw the "gilded" things from the public eye, thereby instantly changing their status and nature.

The great fear of anyone who owns a treasure is that it will be stolen, for the heart of those bewitched by treasure is filled with desire and lust. Haven't you read and seen countless depictions of the clever tricks, con jobs, brilliant robberies executed just so that some bedazzled thug can get his hands on another person's treasure? When a treasure disappears or is stolen, its owner no longer has the pleasure of seeing it, touching it, feeling the shiver that travels from eye to hand to heart when the soul is silenced by admiration.

Modern treasures are often revealed to us by archaeologists. They estimate the "civilizational" value of their discoveries and interpret them for us. They explain the significance of objects from the past, witnesses of unknown identities. And, like Indiana Jones, they fight to ensure that the treasures of humanity end up in and are conserved by museums. Museums are born of the passion of collectors and the "need to show the people the jewels of the Crown." In the late twentieth century, the International Council of Museums (ICOM) defined a museum as "a non-profit-making permanent institution in the service of society and of its development, open to the public, which acquires, conserves, researches, communicates, and exhibits, for purposes of study, education, and enjoyment, the tangible and intangible evidence of people and their environment." Museums are thus giant showcases that exhibit, protect, and conserve the treasures of humanity for posterity.

What makes an object worthy of being designated a treasure? Treasure earns its name by more than its monetary value alone – by the rarity or uniqueness of the thing or the knowledge. In Japan, master crafts persons at the height of their art are considered living treasures, and all Japanese worship and respect them. The nation even provides the needs of such artists, asking only that they devote themselves to their art in order to pass on their skills and knowledge to their apprentices, guiding them to a state of perfect mastery. The majority of modern-day cultures have picked up on this idea and are now interested in and aware of the priceless value of traditional knowledge. Great efforts are being made to catalogue and conserve the intangible heritage of each of these cultures. The value of intangible knowledge is now being recognized because it is the fertile ground for all cultural identity.

---

1. *Civil Code* (France, 1804, art. 716 and ss).

Working tools of the prospector Joe Mann (Chibougamau)

Photograph: Jean-Claude Labrecque, from the documentary *Une rivière imaginaire* (written and directed by Anne Ardouin), ©1993

Museums of history and civilizations ensure the conservation and presentation of cultures and identities by preserving the national collections, including the associated knowledge. This is why objects of material culture are today considered priceless treasures. They are witnesses of our culture, the expression of the knowledge of the generations who preceded us and forged our identity. If art museums acquire, exhibit, and conserve the best examples of human artistic genius, museums of civilization must assume the task and responsibility of collecting and exhibiting works dealing with all aspects of cultural identities. Minority groups as well as lost cultures must be presented and discussed with the same integrity, diligence, and level of interest as majority groups. Museums are therefore entrusted with a sensitive mission: to discover how to reconcile the cultural differences of the ethnic groups and peoples that make up the nation and to ensure that they are represented in a significant way both in the collections and in the exhibitions.

Heritage is more than glorious events and sumptuous objects. Sometimes we have to see and understand an object beyond its form and substance. We must allow an object on display to express all of the knowledge that it contains. We must be able to see the gestures of the person who created and of those who used it – and the love of the person who acquired and treasured it. The beauty and value of any object are in the eye of the beholder. A museum visitor requires a certain distance in order to really understand the jewel put on display for him or her to admire, see, feel, and desire to touch – to really understand the beauty of the treasure that they collectively own. The visitor is invited to read between the lines, as if this is where the truth will finally emerge.

It's up to you to discover the hidden treasures in the biggest treasure chest in the world, during your next visit to a museum.

Terre de labrador

TERRES NOEVVES

Montagnais

Baye de canada

NOVVELLE FRANCE

Armouchicoys

Les vierges

COSTE DE LA FLORIDE

Isles des Esores

TROPICQVE DE CANCER

Isles du Cap de vert

LA LIGNE

ÆQVINOC

MER DV SV

Noeuffues Espaignes

Riuiere des Amazones

LE PERV

Maragnon

Isles cannibal

La France Antartieque

LE BRESIL

TROPICQVE DE CAPRICORNE

MAVRITA

PAR

La serleona

Gold in the Americas: A Voyage Beyond Boundaries

HÉLÈNE DANEAU

Historian and museologist, Project Manager for the exhibition *Gold in the Americas*

Translated by Joan Irving

JUST FIVE CENTURIES AGO, men set sail across the ocean to discover and explore new continents, changing the face of the world forever. They were motivated by a common goal, the same desire: to find the gold and become rich! This metal, which is after all just an ore, profoundly marked the fate of the Americas. Gold seduced and gold destroyed, and it continues to cast a spell on all who touch it.

Is it not said of a deposit of gold that it's a vein leading to the very heart of Earth? Gold ran in rivers and was dug from the bowels of the continent by women and men who almost killed themselves collecting it, so great was their hope of finding their fortunes and their freedom. The peoples of the Americas melted down these fortunes, lost them at the bottom of the sea, gambled and threw them way, but the freedom – the real jackpot – they held on to, and the continents of the New World were seen as lands of liberty. The scars left by the conflicts on the hearts of the Indians, immigrants, slaves, and Métis have not gone away, but the desire for and hope of finding freedom – the unattained riches – signed with one stroke the soul of the Americas.

Like the flame that pushes back encroaching night, the energy that ignites the search for gold caught hold of all those who joined together to make this project happen. Across borders and the vast territory explored in the exhibition, amid a wealth of linguistic and cultural differences, voices and hands united to produce *Gold in the Americas*. The creation of this exhibition was an enriching human and professional experience that resulted in shared knowledge and new friendships. To the conviction that we had made this project possible was added a feeling of belonging. The spirit of the team was far greater than that of the individuals who made it up.

The collaboration and closeness that existed from the outset led to moments of creativity, sensitivity, and emotion that, in turn, left their marks on the exhibition. The energy that you will sense both in visiting the exhibition hall and in turning the pages of this book springs from this exceptional communal spirit. No large-scale exhibition can be produced without the unconditional support of a team of people, and I would like to express my gratitude to all those who worked on *Gold in the Americas* and who have already ensured its success.

Marine map of the Atlantic Ocean
by Pierre de Vaux, 1613
Photograph: AKG Paris

Where will the next El Dorado be?

Watercourse at Cap Tourmente, Quebec
Photograph: Frédérick Bussières

Index

Abd al-Malik, 166
Abitibi, 127, 146, 158; Abitibi-
   Témiscamingue, 160
Aboriginals, 17, 18, 20, 21, 31,
   32, 34, 42, 73, 116, 125, *see also*
   Amerindians
Acosta, José de, 35, 40
Africa, 10, 11, 37, 49-51, 166; East
   Africa, 49; West Africa, 34, 37,
   50, 158; South Africa, 145, 148,
   167; Sub-Saharan Africa, 49
Agnico-Eagle Mines, 160
Aguirre, Lope de, 32, 37
Akkadie, 181
Alaska, 25, 121, 123, 125
Alcantara (chivalric order), 71, 73
Alcántara, Francisco Martin
   de, 60
Alexander the Great, 165
Alexander VI (pope), 18, 63
Alexandria, 49
Allen, Woody, 187
Allumettes (island), 83, 84
Almagro, Diego de, 59
Al Mayo, 121
Alvarado, Juan B., 118
Álvarez Guerrero, Alfonso, 64
Amazonia, 103, 151
America(s), 9, 11, 17, 20-22, 26,
   31, 35, 37, 40, 42, 44, 49, 57,
   63, 64, 73, 83, 127,141, 146-
   148, 157, 177, 184, 201, *see also*
   New World; discovery, 17, 18,
   36, 37, 50, 59, 63, 65, 69, 183;
   North America, 21, 22, 83, 146,
   151, 166; South and Central
   America, 19, 25, 29, 31, 35, 37,
   40, 44, 45, 50, 63, 75, 77, 79, 83,
   85, 91, 94, 98, 140, 166
American River, 115, 119
American Treasury, 169-172
Amerindians (and Indians), 9,
   17-22, 25, 36-38, 43, 58, 59, 63,
   65, 67, 79, 84, 116, 117, 119,
   127, 137, 140, 157, 201, *see also*
   Aboriginals; First Nations, 122
Amsterdam, 20
Amú, William, 140
Andes (Cordillera), 34, 39, 53, 87,
   89, 90, 91, 93, 95, 146, 191, 195
Antilles, 21
Antioquia, 43
Antiquity, 32, 35, 181, 182
Apesteguía, Raúl, 192
Aporoma, 40
Arabia, 49
Aragon (State), 75
Arboleda, Julio, 139
Arévalo, Bernardino de, 64
Argentina, 148
Arizona, 19, 22
Armada (The Invincible), 69-71,
   73, 74

Asia, 18, 50, 75, 77, 79, 170
Atahualpa, 38, 39, 60, 61, 183
Atlantic (Ocean), 18, 23, 35, 37,
   43, 44, 50, 51, 59, 119, 201
Audiencia de Quito, 40
Aurelian Resources Inc., 158
Australia, 146, 168, 174
Austria, 167
Aztecs, 58, 97-101, 116, 182;
   kingdom, 57, 59, 74, 98, 99, 166

Bahamas, 18
Balboa, Vasco Nuñez de, 59
Balsas, 40
Bank of Amsterdam, 20
Bank of England, 21, 167, 168,
   172
Barba, Alonso, 32
Bastian, Adolf, 47
Batán Grande, 90, 191
Beach, Rex, 126
Beauce, 127, 145
Belize, 98
Benté, Carlos Arturo, 139
Bering Sea, 125
Berlin Ethnographic Museum, 47
Beuchot, Mauricio, 66
Bidwell, John, 118
Birch (stream), 121
Bird, Junius, 192
Boèce, 31
Bogotá, 47, 90, 91, 106
Bolivie, 26, 53, 74, 98
Bonanza (stream), 122, 124
Borges, Jorge Luis, 9
Boston, 22
Bourlamaque, 127
Brannan, Samuel, 115, 119
Brazil, 32, 40-44, 50, 63, 75, 149,
   151
Bremen, 47
Bretton Woods (Agreements), 25,
   168, 169, 185
British Treasury, 169, 170
Buenaventura, 137
Byzance, 166

Cadillac, 127, 160
Cahuenga, 118
Cajamarca, 38, 60
Calais, 70, 184
Cali, 137, 138, 140
California, 19; annexation to the
   United States, 22, 116, 118-120;
   Lower California, 117; Upper
   California, 117; gold rush, 22,
   23, 25, 115-120, 127, 168
Camp David, 170
Campbell-Red Lake (mine), 148
Canada, 25, 120, 122, 131, 147,
   148, 157, 158, 169, 177, 198
Canary Islands, 18
Cap Verde Islands, 75

Cap-Rouge, 129
Carabaya, 40
Cardenas, Juan, 32
Caribbean, 9, 18, 21, 37, 39, 50,
   57, 59
Carlin Trend, 146-148
Cartier, Jacques, 84, 127, 129
Caso, Alfonso, 98
Castellanos, Juan de, 46, 79
Castille (State), 63, 67, 75
Castilleja de la Cuesta, 59
Catholic Church, 183
Cauca (river and valley), 40, 104
Central Pacific Railroad, 120
Champagney (Madame de), 73
Champlain, 21
Charlemagne, 166
Charles III, 117
Charles Quint, 58
Chavez, Jerónimo, 41
Chavín (culture), 85, 90
Chiapas, 66
Chiappero, Pierre-Jacques, 155
Chibcha (people), 39, 47
Chicama, 86
Chicomecoatl, 100
Chile, 40, 50, 146, 147-149, 151
Chilkoot, 25, 121-123, 126
Chimú: peuple, 91, 94, 95;
   culture, 191-193
China, 18, 21, 24, 77, 150-153,
   167, 174
Chivor, 77, 79
Chuquiabo, 40
Cieneguilla, 43
Cieza de León, 17, 18
Cihuacoatl, 100
Cipango, 18
Circle City, 121
Ciudad de los Reyes (Lima), 61
Codex Mendoza, 99
Cohen, Hugo, 192
Coloma, 115
Colombia, 32, 37, 39-41, 43,
   47-48, 77, 79, 83, 88, 90-94, 98,
   103,104, 106, 135, 137, 139, 141,
   191, 193-196
Colorado, 117
Columbus, Barthélemy, 65
Columbus, Christopher, 9, 12, 17,
   18, 22, 31, 32, 36, 44, 50, 57, 65,
   77, 97, 183
Columbus, Diego, 65
COMEX (Chicago Mercantile
   Exchange), 171, 173, 175
Company of the One Hundred
   Associates, 21
Compostelle (chivalric order),
   71, 73
Comstock, 120
Connally, John, 170
Conquistador, 17, 18, 32, 35, 44,
   45, 57-60, 73, 75, 91, 103, 197

Constantine I, 183
Constantinople, 18, 166
Cortés, Hernán, 32, 57-59, 73,
   98, 101
Costa Rica, 19, 98
Covarrubias y Leyva, Diego de, 64
Cresques, Abraham, 49
Croesus, 165, 181, 183, 187
Cuauhtémoc, 58
Cuba, 37, 57, 58, 65
Cumaná, 66
Cumberland (river), 127
Cundinamarca, 48
Cuzco, 18, 60, 66, 91, 94

Danube, 150
D'Aquin, Thomas, 63, 64
D'Arcy (gallery), 191
Darien (peninsula), 32, 40
Dawson City, 24, 25, 122, 123,
   125, 126
Del Valle, don Antonio, 117,
   118
De Macas (governor), 36
Denmark, 21, 167
Depestre, René, 11
Deux-Siciles (kingdom), 74
Dionne, Hélène, 9-11
Dominican Republic, 37
Dominicans, 67, 117
Dos Cabezas, 91
Douglas Island, 25
Drake, Francis, 69
Dubai, 175
Duke of Parma, 70
Duran, Fray Diego, 100
Dutch West India Company, 20
Dyea, 123

Eagle's Nest, 155
Ecuador, 36, 37, 60, 91, 92, 94, 95,
   98, 158
Edmonton, 126
Egypt, 49, 152, 165, 181, 184
El Brujo, 91
El Comercio, 191
El Dorado: myth, 18, 22-25,
   37-39, 45, 46, 49, 97, 115, 117,
   119, 197, 202; Klondike mine,
   122, 123
Eleonore (deposit), 158, 161
Elizabeth I, 69
Ell (lake), 161
El Limbo, 138
El Salvador, 98
Emmerich, André, 192
England, 19, 21, 22, 69, 167, 168,
   185; English Crown, 21, 184, *see
   also* Great-Britain
English Channel, 69
Esmeraldas, 36
Ethiopia, 49
Europe, 9, 17, 18, 20, 22, 23, 34,

Palos, 31
Panama, 18, 23, 32, 59, 60, 98,
    137, 182
Papaloapan, 40
Paraguay, 37
Parthenon, 183
Pasadena Union, 117
Pasca, 48
Pascalis, 127
Perrenoto, Tomás, 73
Persia, 165
Peru, 18-20, 37, 41, 50, 60, 61, 66,
    74, 77, 90, 107, 147, 148, 151,
    193, 195; art and goldsmithery,
    85, 86, 89, 91, 93, 96, 98, 108,
    183, 191, 196; north coast, 39,
    86, 89-91, 191
Philadelphia, 166
Philipe II, 66, 69-71, 73
Philippines, 25, 77
Pizarro, Gonzalo, 59, 60
Pizarro, Hernando, 60, 61
Pizarro, Juan, 60
Pizzaro, Francisco, 32, 38, 39, 57,
    59-61, 183
Placer County, 155
Placerita Caceta, 119
Placerita Canyon, 117, 118, 120
Plantes (rivers), 127
Polk, James, 119
Pontificia Universidad Católica,
    195
Popón, 103
Portland Bill, 69
Portland, 25
Portugal, 18, 20, 44, 49, 63, 69,
    73-75; Portuguese Crown, 43,
    49, 73
Potosí, 74, 77
Printemps des peuples, 23
Promontory Point, 120
Puebla de los Angeles, 79
Puerto Rico, 37

Qosqo (Cuzco), 60
Quebec: province, 127, 129, 145,
    147, 158, 161, 202; Quebec
    City, 177
Quesada, Gonzalo Jiménez de, 46
Quesada, Hernan Pérez de, 46
Quetzalcoatl, 99, 100
Quimbaya (people), 90, 92, 104
Quito, 36, 39, 46

Rabbit (stream), 122
Ramos Ruiz, Gonzalo, 47
Rancho San Francisco, 117
Real de Todos Santos, 41
Reconquista, 18
Red Lake, 148, 157
Reinel, Pedro and Jorge, 35
Renaissance, 34, 35, 71, 183-187
Republic of Indians, 67
Restall, Matthew, 57
Rio de Janeiro, 51
Rio de la Plata, 19
Río Nechi, 41
Río Timbiqui, 135, 137, 140
Roberto (deposit), 161
Roberval, Jean-François de La
    Rocque de, 129

Rogers, Woodes, 138
Rome, 65, 165
Roosevelt, Frank D., 168, 185
Romania, 150
Rousseau, Jean-Jacques, 187
Rueff, Jacques, 170
Russia, 21, 117, 167, 170

Sabara, 43
Sacramento, 24, 115, 119
Sadiola, 158, 159
Saguenay (and Saguenay–Lac-
    Saint-Jean), 84
Sahagún, Fray Bernardino de,
    98, 100
Sahara, 49, *see also* Africa
Saija (river), 136, 140
Salamanca, 57, 66, 67
Salazar Sola, Carmen, 53
San Bueno Ventura, 117
San Emidio Canyon, 117
San Feliciano, 117
San Fernando, 117, 118
San Francisco, 23, 24, 115, 118-120
San Francisco, Javier, 118
San Francisquito, 117
San Joao de Mina (Saint John of
    the Mines), 50
San Luis Potosí, 43
San Pedro, 43
Santa Bárbara, 137
Santa Clarita, 117
Santa María, 137
Santo Domingo, 65
Sao Paulo, 40
Scandinavia, 167
Schubnel, Henri-Jean, 155
Scotland, 21, Mine, 170
Seattle, 25, 122
Second World War, 169, 170
Seler, Edouard, 47
Sepúlveda, Antonio de, 46
Sepúlveda, Juan Ginés de, 64, 67
Serra, Junípero, 117
Serra Pelada, 26
Service, Robert, 126
Seville, 19, 57, 58, 65, 66, 74, 75
Shangai Gold Exchange, 175
Shöerwel, Gabriela, 195
Siberia, 89
Sicán: culture, 88, 90, 91, 191-
    193; people, 91, 92, 94, 95
Siecha (lagoon), 47
Sierra Leone, 51
Sierra Nevada de Santa Marta, 103
Sierra Nevada, 116, 120
Silver: exploitation, 40, 44, 49,
    57, 73, 79, 90, 120; metal, 145,
    150, 152; objects, 47, 66, 74, 77,
    79, 85, 90-93, 98, 191, 194, 196;
    compared to gold, 24, 77, 92,
    93, 165, 167, 181
Simón, Fray Pedro, 46
Sinai (mount), 182
Sinù (river), 39
Sipán, 196
Sixtymile (river), 121
Skagway, 123
Slavery (and slaves), 11, 19, 20,
    39, 40, 43, 50, 51, 64, 66, 100,
    119, 136, 137, 140, 141, 201

Smith, Adam, 185
Smithsonian Institute, 171
Soapy Smith, 123
Société des lingots d'or, 23
Solomon, 32, 182
Solon, 181
Sonora, 40, 116, 118
Soto, Fray Domingo de, 64, 67
Spain, 18-20, 31, 36, 37, 57, 59-
    66, 69, 70, 73-75, 77, 91, 127,
    141, 166, 167; Spanish Crown,
    18-20, 36, 43, 46, 57, 63-65, 67,
    75, 79
Spinola, Fabricio, 70
Spondylus princeps (shell), 94, 95
St. Lawrence (valley and river),
    84, 129
Stanford, Leland, 120
Stenuit, Robert, 70, 73
Stewart (river), 121
Sudan, 49
Sun: worship, 32, 34, 37, 46, 92,
    93, 98, 100, 182, 183; temple,
    18, 39, 94, 107
Superior (lake), 83, 84
Sutter, John, 115, 118-120
Sweden, 21, 167
Switzerland, 167, 169, 172

Tabasco, 58
Tairona, 91, 92, 103
Tayta Muki, 53, *see also* Muki
Tenochtitlán, *see* Mexico City
Texas, 19, 22, 119
Timmins, 146, 151, 157
Tio, 53
Titicaca (lake), 39, 91
Tiwanaku, 39, 91
Tlacaltecs, 58
Tlaxcala, 58
Toscanelli, Paolo, 18
Total Group, 155
Totonaques, 58
Tutankhmon, 145
Toxcatl, 58
Treaty of Alcaçovas, 75
Treaty of Guadalupe-Hidalgo,
    115, 118
Treaty of Tordesillas, 18, 63, 64,
    75
Treaty of Versailles, 168
Tr'ondëk Hwech'in, 125
Trujillo, 59, 107, 108
Tucoya (Indian), 117
Tudor, Mary, 69
Túmbez, 59
Ulster Museum, 70, 74
Union Pacific Railroad, 120
United Provinces, 20, 21
United States, 22, 24, 115, 118-
    120, 122, 127, 141, 146, 167-
    171, 175, 186; gold stocks, 171,
    174, 186
Ursua, Pedro de, 37
Uwa (people), 103

Val-d'Or, 127, 147, 157
Vancouver, 25
Vatican, 65
Velasquez, Diego de, 57
Venezuela, 37, 73

Vera Paz, 66
Veracruz, 58, 99, 101
Veraga, 40
Vicús (culture), 90
Vietnam, 170
Virgil, 35
Virginia Gold Mines, 158, 161
Virginia, 21
Vitoria, Fray Francisco de, 64, 67
Vives, Luis, 64
Volcker, Paul, 172
Von Hutten, Felipe, 37

Wales, 167
Wari, 91
Wemindji, 161
West Indies, 9, 18, 20, 35, 57, 64,
    75, 77
West, Robert, 140
White, Harry, 169
Wight (Isle of), 70
Wimmer, Peter, 115
World Bank, 25, 137
World Gold Council, 152, 186
World Trade Center, 172

Xipe Totec, 98-100
Xiuhtecuhtli, 101
Xochiquetzal, 100

Yananocha, 148
Yopico, 98
Young, Ewing, 117
Yucatan, 58
Yukon, 25, 32, 121-126, 128, 168;
    river, 121-123

Zambo (republic), 36
Zuñiga, Daniela, 137
Zurbarán, 77

Le Vaisseau D'Or

C'était un grand vaisseau taillé dans l'or massif.
Ses mâts touchaient l'azur sur des mers inconnues
La cyprine d'amour cheveux épars chairs nues
S'étalait à sa proue au soleil excessif

Mais il vint une nuit frapper le grand écueil
Dans l'océan trompeur où chantait la Cyrène
Et le naufrage horrible inclina sa carène
Aux profondeurs du gouffre immuable cercueil!

Ce fut un vaisseau d'or dont les flancs diaphanes
Révélaient des trésors ainsi que les marins profanes
Dégoût Haine et Névrose ont entre eux disputé

Que reste-t'il de lui dans la tempête brève
Qu'est devenu mon cœur navire déserté
Hélas! il a sombré dans l'abîme du rêve.

Émile Nelligan, I.W.

4 Mars 1912.

◀ *Le Vaisseau d'or*
Handwritten poem by Émile Nelligan, 1899
Photograph: Bruno Rodi

COMPOSED IN MINION BODY 11 AND IN ARGO
ACCORDING TO A LAYOUT DESIGN BY PIERRE-LOUIS CAUCHON
THIS BOOK WAS PRINTED IN APRIL 2008
ON HORIZON GLOSS 200 M
UNDER THE ATTENTIVE EYE OF YVON BÉGIN AT PRESSES LITHOCHIC IN QUEBEC CITY
AND BOUND AT ATELIERS MULTI-RELIURE S.F.
AND THIS FOR GILLES HERMAN
PUBLISHER AT LES ÉDITIONS SEPTENTRION